Documents, Contracts, and Worksheets for Home Builders

More Construction Books from McGraw-Hill

BIANCHINA • *Forms and Documents for the Builder*

BOLT • *Roofing the Right Way, 3/e*

BYNUM AND RUBINO • *Handbook of Alternative Materials in Residential Construction*

CLARK • *Retrofitting for Energy Conservation*

DOMEL • *Basic Engineering Calculations for Contractors*

FRECHETTE • *Accessible Housing*

GERHART • *Everyday Math for the Building Trades*

HACKER • *Residential Steel Design and Construction*

HARRIS • *Noise Control Manual for Residential Buildings*

HUTCHINGS • *National Building Codes Handbook*

JAHN AND DETTENMAIER • *Offsite Construction*

KOREJWO • *Bathroom Installation, Design and Remodeling*

KOREJWO • *Kitchen Installation, Design and Remodeling*

MILLER AND BAKER • *Carpentry and Construction, 2/e*

PHILBIN • *The Illustrated Dictionary of Building Terms*

PHILBIN • *Painting, Staining and Refinishing*

POWERS • *Bathrooms: Professional's Illustrated Design and Remodeling Guide*

POWERS • *Kitchens: Professional's Illustrated Design and Remodeling Guide*

SCHARFF AND THE STAFF OF ROOFER MAGAZINE • *Roofing Handbook*

SCHARFF AND THE STAFF OF WALLS & CEILINGS MAGAZINE • *Drywall Construction Handbook*

SHUSTER • *Structural Steel Fabrication Practices*

SKIMIN • *The Technician's Guide to HVAC Systems*

VERNON • *Professional Surveyor's Manual*

WOODSON • *Be a Successful Building Contractor, 2/e*

Dodge Cost Books from McGraw-Hill

MARSHALL AND SWIFT • *Dodge Unit Cost Book*

MARSHALL AND SWIFT • *Dodge Repair and Remodel Cost Book*

MARSHALL AND SWIFT • *Dodge Heavy Construction Unit Cost Book*

MARSHALL AND SWIFT • *Dodge Electrical Cost Book*

Documents, Contracts, and Worksheets for Home Builders

Al Trellis
Home Builders Network, LLC

McGraw-Hill

New York   San Francisco   Washington, D.C.   Auckland   Bogotá
Caracas   Lisbon   London   Madrid   Mexico City   Milan
Montreal   New Delhi   San Juan   Singapore
Sydney   Tokyo   Toronto

Library of Congress Cataloging-in-Publication Data

Trellis, Alan R.
    Documents, contracts, and worksheets for home builders / Alan R. Trellis.
        p.   cm.
    Includes
    ISBN 0-07-065355-0 (hardcover). — ISBN 0-07-065354-2 (pbk.)
        1. Contractors—Forms.   2. Construction industry—Forms.
    3. Building trades—Forms.   4. Construction contracts—Forms.
    I. Title.
        HD9715.A2T7   1998
        651′.29′02469—dc21   98-18202
                                                                                      CIP

McGraw-Hill

A Division of The McGraw·Hill Companies

1 2 3 4 5 6 7 8 9 0   KGP/KGP   9 0 3 2 1 0 9 8

P/N 0-07-913769-5 (PBK)   P/N 0-07-913768-7 (HC)
Part of   Part of
ISBN 0-07-065354-2   ISBN 0-07-065355-0

*The sponsoring editor for this book was Zoe Foundotos, the editing
supervisor was Frank Kotowski, Jr., and the production supervisor
was Sherri Souffrance. It was set in Arial by Brad Trellis.*

Printed and bound by Quebecor/Kingsport.

McGraw-Hill books are available at special quantity discounts to use as
premiums and sales promotions, or for use in corporate training programs. For more information, please write to the Director of Special
Sales, McGraw-Hill, 11 West 19th Street, New York, NY 10011. Or contact your local bookstore.

 This book is printed on recycled, acid-free paper containing
a minimum of 50% recycled, de-inked fiber.

Contents

Preface ix
Acknowledgements xv

BID PACKAGE 1
Client Contact Form and Specifications Checklist 3
Bid Letter 9
Summary Estimating Form 13
Estimating Assumptions, Recommendations, and Bid Analysis 19
Estimating Assumptions/Recommendations 21
Bid Summary Analysis 23
Professional Services Agreement 25
Building Permit Worksheet 29
Five Reasons to Get Bids 31
Bid Organization Form 33
Bid Cover Sheet 35
Standard Suubcontractor Conditions for Bidding 37
Sample Bid Forms
   Masonry 39
   Rough Carpentry 41
   Plumbing 43
   Electrical 45
   Heating and Air Conditioning 49

CONSTRUCTION PACKAGE 51
Construction Schedule Form 53
22 Week Construction Schedule 57
30 Week Construction Schedule 61
Subcontracts 65
Trim Carpentry Subcontract 67
Frame Carpentry Subcontract 69
Draw Schedules 71
Construction Draw Schedule 73
Functions 75
Functions—ABC Builders 77
Checklists 81
Initial Checklist (Questions/Comments) 83
Framing Checklist 85
Plan Review Checklist 87
Pre-construction Carpentry Meeting Form 91
Pre-Drywall Checklist 95
Sample Order Form 101
   General 103
   Masonry Block 105
   Structural Steel 107
   Framing Lumber 109
   Roof Lumber 111
   Exterior Windows and Doors 113
   Interior Trim 115

CONTRACT PACKAGE 119
Contract and Associated Documents 121
Fixed Sum Contract 123
Cost Plus Fixed Contract 131
Allowance Schedule 139
Specifications 145
Charges/Prices 159
Design Contract 163
Policy for Owners Using Their Own Subcontractors and Suppliers 177
Regulatory Checklist 181
Request for Architectural Committee Approval 183
General Information Worksheet for Architectural Committee Approval 185
Exterior Material Worksheet for Architectural Committee Approval 187
Application for Utility Service 189
Building Permit Worksheet 191

CLIENT INFORMATION PACKAGE 193
Client Package 195
Client Notebook 197
Realtor/Clent Registration Form 199
Request for Information 201
Prospect Card 203
Referral Card 205
Contract Acceptance Letter 207
Scheduling Information 209
Client Selections 211
Client Selection Schedule 213
Client Selection Form 215
Contract List for House Selections 223
Lighting Fixture Selections 225
Change Orders 227
Quality Control and Material Characteristics 231
Walk-Through Inspection 233
Walk-Through Inspection Information Checklist 235
Service Call List 241
Kitchen and Bathroom Design Worksheets 243
  Cabinets 245
  Appliances 247
  Countertops 249
  Mechanical layout 251
Bathroom Design Worksheets
  Vanities, Countertops, Accessories 253
  Plumbing Fixtures 255
  Ceramic Tile 257

CUSTOMER SERVICE PACKAGE   259
Pre-Contruction Conference Agenda   261
Warranty Forms   263
Welcome to Warranty   265
Year End Letter   267
Phone Log   269
Warranty Service Request Form   271
One Time Repair Request   273
Warranty Work Order   275
Work Order Log   277

OFFICE MANAGEMENT PACKAGE   281
Fax Cover Sheet   283
Transmittal Sheet   287
Sales Office Checklist   289
Marketing Materials Checklist   291

PROFIT MANAGEMENT PACKAGE   295
Month End Reporting   297
Income Statement   299
Balance Sheet   303
Ratio Analysis   307
Job Cost Analysis   309
Cash Flow Forecasting   311
Sample Builder Monthly Cash Flow Projection   313
Yearly Marketing Budget   317
Weekly Exception Report   321
Weekly Status Report   323
Problem/Solution Report   325
Year End Planning Checklist   329
Strategic Planning Form   331
Competitive Shopping Guide   337

Preface

THE IMPORTANCE OF MANAGING INFORMATION

Builders sometime think that managing the building business is about managing construction and managing people. But it's much more than that. Like any other business, it's about managing information: information about clients, design, schedules, and money. If you're smart about managing that information, you'll find that building houses can be more fun and more profitable.

As an example: two builders build similar houses – they have the same number of square feet, same number of bedrooms, and are located in similar neighborhoods. But one of these houses has small dark rooms, poor circulation, poor detailing and inappropriate materials. The other uses volume spaces, light-filled rooms, and good attention to detail. One house sells quickly at a higher price – the other sits on the market, if it sells at all. What's the difference? Information, in this case design information. Knowing what to build, what people want in a home and what they are willing to pay, makes all the difference between success and failure and between profit and loss.

Suppose that these two builders built the identical house (same design and client information). But, one builder has an information system that tells him what costs he's occurring on a timely basis, so that he can compensate for mistakes and cost overruns during the construction process. He also controls his scheduling information so that he can plan ahead and avoid or compensate for construction delays as they occur. The other builder runs his business by the seat of his pants. At the end of the job, which builder is going to have more profit? Again, the difference is information.

Being smart about managing information isn't obvious. It's easy to get overwhelmed by the data being collected, and lose sight of the forest for the trees. If you don't have a systematic method of collecting the information before you need it, you often find yourself scurrying around at the last minute, making phone calls, and contacting clients to obtain information, with a little foresight, could have been collected with much less effort.

There is a big difference between data and information. Data is the raw unprocessed bits and pieces of information, before it's put into a usable format. A phone number scribbled on a piece of paper is data. When you combine it with name, address, and housing requirements to form a client profile, then it's information. Information tells you what that data means in terms and tasks and profits. Information management consists of collecting the right data, in a usable format, and understanding the significance of the information collected. For example, having the phone number of a potential client does little good unless it's tied to a systematic method of calling those potential clients to convert them into buyers.

Here are five rules for smart information management:
1. Keep it simple
2. Keep it organized
3. Keep it easy to use
4. Keep it up to date
5. Keep it profit driven

KEEP IT SIMPLE

There are two mistakes you can make in collecting information: (1) Collecting too little, or the wrong information, or (2) collecting too MUCH information. It takes time and resources to manage information, and the more time we spend collecting it, the less time we have for other activities. This is especially true in larger organizations, where reports often take on a life that far exceeds their usefulness. So the first step in keeping it simple is to eliminate unneeded and useless forms. Go through all the forms you're currently using, and ask yourself this question: If I didn't collect this data, what difference would it make? If the answer is none, then eliminate the form. If the data is important, but the form is too complicated, redesign it and simplify it.

The next step is to decide what information you NEED that you aren't collecting at the present time. This information should be collected before you need it, rather than hurrying to find it at the last moment.

The many forms included in this book can give you valuable insights into what information can be collected. Remember, however, that not all the forms in this book will pertain to you or your operations. In the interest of keeping it simple, select only the forms you will actually need, and implement them. You can always add more forms later if you decide you need them.

Whatever you do, don't create a headache for yourself by creating an information management system so complex and cumbersome that it defeats the purpose of smart management. The system is there to serve you. You're not there to serve the system.

KEEP IT ORGANIZED

Nothing is more frustrating than searching high and low for information that you know you have, but can't find. Misplaces files, forms filed in the wrong sequence, or information scribbled on scraps of paper and then stuck somewhere you'll remember (but never do) – these are the bane of good information management. You should be able to retrieve the information you need in a few minutes without going to a great deal of effort.

There is a simple solution: keep all information filed by job in one place. We use a separate 3-ring binder for each house we build with the book organized using tabs into 8 separate sections of information. To find a particular form, you simply turn to the relevant section, and thumb through the forms until you find it. All forms are filed in reverse chronological order with the most recent date on top. That way, if there are changes, you're dealing with the most recent information. Since all the forms are punched and inserted in the binder, nothing can get out of order.

The 3-ring binder never leaves the office. That way you avoid the catastrophe of losing the critical information you need to manage the job. Copies of the relevant forms can be made to take into the field, so they are available where needed.

Even if you keep the bulk of your information in a computer database system, it's important to have paper backup of critical documents. Most builders now use computerized accounting system to track expenses by job costs. They may also use computerized scheduling and estimating programs. All of the programs can produce reports that are critical to the management of a particular project. The paper output of the report should also be filed in the binders. Paper input such as change orders, invoices, bid estimates from contractors and suppliers, etc., become input into these programs and should be filed in the project binder.

The data in these computer programs is extremely valuable and should be backed up electronically on a daily basis. Backup disks should be stored off the premises or in a fireproof safe. Nothing is worse than coming to the office one day and discovering that your computer has been stolen, or your hard drive has crashed along with all your valuable information. You can always replace a computer, but information is often irreplaceable.

KEEP IT EASY TO USE
By keeping your information simple and organized, you also help keep it easy to use. The term "user-friendly" is usually applied to computer programs, but it can also apply to any management system. Computers can be powerful tools for managing information and performing calculations. However, you should only computerize where it provides a clear advantage. A good simple manual system that is used on a daily basis is much better than a complex computer system that is difficult to learn and implement. We've talked to lots of builders over the years who have spent thousands of dollars on complex integrated builder management programs, and then spent weeks or months trying to learn them, only to finally give up and leave the software in a box gathering dust. In evaluating any computer management software, make sure that ease of use and ease of learning are high on your list of criteria.

Where the computer can make calculations on the data, such as an accounting system, computerizing makes sense. Where the data is only entered once, and then referred to manually, it may not make sense. Some of the newer integrated computer management systems actually go so far as to provide custom data input forms that your field personnel can fill out manually and fax directly into the computer. The computer reads the forms and enters the data -- no rekeying of information. Based on these forms, it will then generate custom task lists for your personnel. Your personnel can interface with your computer system without becoming computer jockeys, or carrying around a laptop computer. The system is able to use technology to take advantage of the power of the computer, and to keep the interface simple for the user.

KEEP IT UP TO DATE
Information is only useful is it is available in a timely manner. The time to find out about a cost overrun in the framing stage is during the framing stage, when you can do something to compensate for the overrun, not after the home is completed. To keep your information up to date, you must collect the information before its needed, and enter all data on a daily basis.

At the VERY least, all costs and problems should be reviewed job-by-job on a monthly basis. If your system can generate exception reports and problem reports, these should be reviewed, at minimum, on a weekly basis.

KEEP IT PROFIT DRIVEN
You would think that it would be a given that all management systems are driven by the need to control and increase profits. Sadly, this is not always the case. Often, management systems look at one aspect of the business, (expenses, schedules, customer service, etc.) and try to control for only that one aspect, without regard for its effects on profits.

Marketing, for example, is usually treated as an expense. The less you spend on marketing, the more is left for profit. Yet any businessman knows that you have to have marketing in order to generate sales and without sales you have no profits.

A profit driven system basically sets parameters for actions and then compares actual events with these anticipated events. If they are the same (or within a small range of variation) it ignores them. If there is a discrepancy (sales are below expectations, or costs on a given project exceed budget) it immediately points this out. This is known as management by exception. You spend your time and energy on those areas that have become problems, rather than getting lost in all the details of the information you collect.

A variation of management by exception is called management by objective. You set goals for each of your business areas (sales, construction, customer service, etc.) and then evaluate them on a consistent basis based on how they are meeting those goals. The goals themselves, however, need to be profit-driven. Customer service, for example, can have a negative effect on profits by spending too much money trying to meet customers unrealistic expectations. It can also have a negative effect by spending too little, and not creating the customer satisfaction that leads to referrals, and increased sales.

The point we're trying to make is that you can't focus exclusively on one area of the business (expenses for example) if you want to increase profits. You have to understand the dynamics of how all the aspects work together to increae profits.

THE CLIENT DATA BOOK
In the section "Keep it Organized", we talked about creating a separate 3-ring binder for each project. This system works best if you're building a limited number of houses at any one time. If you're highly organized and a filing system works for you, that's also fine. In the custom building business, however, the amount of paperwork required for a house would fill several file folders by the time the job is completed, and would be cumbersome for finding a particular piece of information. We use tabs to subdivide the information which makes finding the relevant documents easier and faster.

Sections a client data book should contain
- Permits
- Contracts/specifications
- Costs and Change Orders
- Subcontractors
- Orders
- Draws/Financing
- Monthly schedule update
- Warranty/Customer Service

You may find that you want to add other sections to your client data book.

HOW TO USE THIS BOOK
This book is only valuable to you if you use it, and begin applying the information to manage your business smarter. Use the following steps to implement this system:

1. **Select the forms you need to manage your profits**. This book contains many different forms you can use to organize your information management. Some of them may apply to your business, some may not. As you go through the book you may be tempted to begin adding new forms and procedures to your operation. Remember the principle of "Keep it simple" and only add those that provide information crucial to your operation.
2. **Modify the forms for your particular needs**. To make this easy for you, all the forms are provided in electronic form on diskette. Simply find the form, make the changes you need (fonts, logos, margins, specific printer, etc.), save it as whatever name you choose (Jones 1 for example) and print it out.
3. **Get organized**. Collecting more information does you no good if it isn't organized and easy to use. Create your Client Data books, and file information on a timely basis.

4. **Get started**. Nothing happens until you make the commitment and start doing it. Start small, and make minor improvements, rather than trying to create a wholesale reorganization that can put your organization into chaos. Do one step at a time, and keep improving until the system functions the way you want it to.

By applying these concepts in a systematic and incremental manner, you can begin to manage the information your business needs to grow and succeed. And that's what smart information management is all about.

Home Builders Network, LLC

Acknowledgements

We would like to thank all the people who contributed to this work, especially Carol Smith, who created the Customer Service forms and procedures, and Steve Maltzman, CPA, who created the financial management forms. Both of these people are leaders in their respective areas of expertise and have provided valuable assistance to the building industry.

We would also like to thank Brad Trellis of Home Builders Network, LLC for the hours he spent reviewing and formatting the forms, preparing them for publication.

Finally, we'd like to thank all the builders we've talked to and worked with over the years, whose problems and concerns provided the impetus for the creation of these forms and documents. Without their input, this publication would not exist.

Documents, Contracts, and Worksheets for Home Builders

BID PACKAGE

Client Contact Form and Specifications Checklist

The Client Contact Form is a way to present an organized image at the first meeting with the client. It ensures that all the appropriate questions will be asked of the client and provides a mechanism for the orderly recording of the information. Properly prepared, this contact form will serve as the preliminary specifications from which the estimate and the final specifications will be prepared.

For the contact form to be effective, you must have previously determined default selections, that is, those items which you include as "standard" features in your homes. An example might be a Kohler elongated bowl toilet in a choice of standard colors. When you meet with the client, you display a picture of the toilet and say: "This is our standard toilet and it is available in your choice of standard colors. If you want it in any of the premium colors (here you give the examples of several premium colors), there would be a slight extra charge. Additionally, if you want any other toilet, it would be either less expensive or more expensive, depending upon which you select." At this point the client should agree that "That toilet's fine," or they may ask to see what else is available.

For this meeting to be efficient, and for the client contact form to be filled out quickly, you must have the available materials so you can say to the client, "Let me show you several other toilets." If they like one, you should be able to give them, if not the exact price, at least an approximate price differential.

It is important to be able to do this on many of the key items, although it is not necessary to do it with everything. For example, if the client wants something very special, you may say, "This is a very special item, and if you want something like that, I will price it in the estimate. However, I don't have the price available right now."

Every good client contact form should have the following information:

1. All personal information about the potential clients, including their names, home and work phone numbers, mailing address, etc., so that you can contact them during the estimating and preliminary phases.

2. A thorough, step-by-step breakdown of the major systems and selections. This analysis must begin with questions about the lot (flat, steep, wooded, clear) and progress systematically through the house. From wall thickness (2x4 or 2x6) to roof type, to number of recessed lights, the more detailed and complete the questionnaire, the greater the accuracy of the estimate. Additionally, fewer areas left undefined reduce the chances of misunderstandings and unfulfilled expectations.

BUSINESS BUILDER

This comprehensive approach will demonstrate above all else to the clients that they are dealing with an organized, professional builder. Convincing your clients early on of your competence is one of the most important aspects of a successful marketing program. Unless your competitors can do the same, you will have an overwhelming advantage.

ABC BUILDERS
123 Main Street
Hometown, MD 21234
(301) 231-1234

Name:_____

Address: _____

Telephone: (h) (_____)_____-_____ (w) (_____)_____-_____

1. Approximate starting date _____

		YES	NO
2.	Is lot surveyed?	_____	_____
	Grading Plan required?	_____	_____
	Permit Relocation Required?	_____	_____
	Any Covenants?	_____	_____

3. Is lot wooded? If so, how much? _____

4. Septic _____
    or Sewer _____
    If septic - disposal?   YES _____   NO _____

5. Well _____
    or Water _____

6. Any demolition?   YES _____   NO _____

7. WALLS: 2x4_____ 2x6_____
    or Other _____
    FLOOR JOIST: 2x10_____ 2x12_____
    TJ1_____

8. Trim description: Traditional_____
                        Contemporary _____

9. Main Stairs: Oak _____ Pine _____ Stain _____   Carpet _____
    2ND Stairs: Oak _____ Pine _____ Stain _____   Carpet _____
    Bsmt. Stairs: Oak _____ Pine _____ Stain _____   Carpet _____

10. Int. Doors: Paint _____ Stain _____ Flush _____   6 panel _____

11. Ext. Doors: Front - Steel_____ Wood_____Other _____
    Bsmt. - SGD
    French _____ Prado _____ Atrium _____
    Other – Steel _____ Wood _____ Other _____

12. Windows: Anderson _____ Pella _____ Alum. _____ Other _____

13. Roof: STD _____ Horizon _____ Timberline _____ Wood _____

14. Plumbing: Water Heaters/Size: One (1) _____ or Two (2) _____
    Toilets: Standard _____ Special _____
    Sinks: Standard _____ Special _____
    Whirlpool: YES _____ NO _____
      Description: _____

15. HVAC Specialties:
    Number units _____ heat pump _____ oil _____ other _____
    Comments:_____

	YES	NO
Humidifiers:	_____	_____
Electronic Filters:	_____	_____
Laundry Chute:	_____	_____

16. Electrical: Number recessed: _____
    Estimated Service: 200 AMP _____ 300 AMP _____ 400 AMP _____
    Grounds Lighting: YES _____ NO _____
    Special Requests/requirements: _____

17. Insulation:
    Walls: R-13 _____ R-19 _____ Other _____
    Ceiling: R-30 _____ R-38 _____ Other _____
    Crawl: _____
    Basement: Walls _____ Ceiling _____

18. Drywall: Garage; common wall & ceiling only: _____
    Entire garage: _____

19. Paint/Stain: Unusual Requirements _____

20. Siding: Unusual Requirements _____

21. Finish Driveway: Approx. length _____
    Stone _____ Asphalt _____

22. Culvert/Apron: Culvert required? YES _____ NO _____
    Apron required?   YES _____ NO _____

23. Garage Doors: Single Doors _____ Double Doors _____
    Materials: _____
    Style: Flush _____ Paneled _____
    Number of openers _____

24.　Fireplaces:

Location	Type	Hearth	Raised Hearth	Flush Hearth Material	Hearth Material	Profile Fan	Doors	Gas Logs
_____	____	_____	_____	_____	_____	_____	_____	_____
_____	____	_____	_____	_____	_____	_____	_____	_____
_____	____	_____	_____	_____	_____	_____	_____	_____
_____	____	_____	_____	_____	_____	_____	_____	_____
_____	____	_____	_____	_____	_____	_____	_____	_____
_____	____	_____	_____	_____	_____	_____	_____	_____

25.　Countertops: Kitchen:　Laminate _____

　　　　　　　　　　　　　　　　Tile _____

　　　　　　　　　　　　　　　　Corian _____

　　　　　　　　　　　　　　　　Other _____

　　　　　　　　　Baths:　Laminate _____

　　　　　　　　　　　　　　Tile _____

　　　　　　　　　　　　　　Cultured Marble _____

　　　　　　　　　　　　　　Other _____

26.　Shower Door: Quantity - Standard _____

　　　　　　　　　　　　　　Special _____

27.　Appliances Kitchen: (check if included)

_____ Refrigerator

　　　　　　　　　_____　　_____　　_____　　_____
　　　　　　　　　19/22cf　24cf　26/27cf　Subzero (1or2)

_____ Range　　Single oven _____

　　　　　　　　Double oven _____

_____ Cooktop - downdraft　4B _____　6B _____

　　　　Other _____

_____ Wall oven -　　　single _____

　　　　　　　　　　　double _____

_____ Microwave

_____ Disposal - above average _____　best _____

_____ Dishwasher - above average _____　best _____

_____ Washer - Details _____

_____ Dryer – Details _____

Appliances - Other _____

_____

28.　Carpet:　Dollar value of allowance (in place) - $/yd _____

29.　Hardwood:　Locations _____

_____

　　　　Strip oak _____

　　　　Other _____

30.　Vinyl: Locations: _____

Dollar value of allowance (in place) - $140 _____

31. Ceramic Tile/Marble:
Kitchen: _____
Foyer: _____
Master Bath: _____
Other: _____
Other: _____
Other: _____
Other: _____
Other: _____

32. Intercom: YES _____ NO _____

33. Decks: Approx. Sq. Ft. _____
Special requirements: _____
Porches: Approx. Sq. Ft. _____
Included in Contract? _____

Special Features:	YES	NO
Security System	_____	_____
Wall Hung Ironing Board	_____	_____
Central Vacuum System	_____	_____
Greenhouse	_____	_____
Entry Piers, Walls, Gates	_____	_____
Safe(s)	_____	_____
Cedar Closet	_____	_____

This questionnaire has been completed by:

_____   _____
Client   Date

Received by:

_____   _____
ABC Builders   Date

Bid Letter

After you meet with the clients, discuss their needs, and determine preliminary specifications you will use to estimate their home, it is important to send a written letter with the estimate. This letter should include the following:

1. A clear definition of the criteria used in the estimate.

2. A statement that this price is contingent upon final preparations of plans, specifications and contract documents. State specifically that certain details may not be evident in the preliminary design.

3. A period of time for which this price estimate is valid. This period of time should include deadlines for both the signing of a contract and commencement of construction. The consequences of a client-ordered delay should be clearly defined.

Finally, if this contract price if for house only, it should specifically state that, or if land financing costs are included or excluded, these also should be specifically stated. Additionally, a statement to the effect that the cost estimate is provided solely as a service to the client and should be kept confidential should be included. Such a statement will prove helpful in inducing many clients not to show the estimate to other builders.

Any additional information which can be placed into the letter, specifically relating to the clients' home, is also beneficial. Cost-cutting recommendations, specification alterations, or general observations all show the clients that you have taken the time and energy to diligently pursue the estimating process on their home. This detailed review demonstrates the type of professionalism which hopefully will set you apart from competing builders.

ABC BUILDERS
123 Main Street
Hometown, MD 21234
(301) 231-1234

Sample

Date

Inside Address

Dear _____:

We are pleased to provide you with the preliminary estimate shown below. This estimate was developed from the plan/sketch/and drawings provided as well as the Client Contact Sheet attached. Obviously, this price is contingent on preparation of final plans, specifications, and clarification of details not evident in the preliminary design phase.

Additionally, the price is valid for a contract signed within sixty days (60) of this date, and a construction starting date as specified below. Client-induced delays beyond this time may result in a cost increase. All prices are quoted exclusive of land or financing costs.

Price Quotation:

Date of This Estimate: _____

Estimated Construction Start: _____

Contract Price: _____, includes the following allowances:

The attached Cost Estimate is the property of ABC BUILDERS and is provided as a service for your personal use. Please keep this document confidential. Thank you.

Sincerely,

Joe Q. Law
President/ABC Builders

ART/pat

attachment

Summary Estimating Form

The review of the estimating analysis is often called the summary estimate or master estimating sheet. This sheet consists of a list of all the appropriate cost categories, along with the estimated cost of that category and the estimated cost of the project. Every summary cost sheet, whether shown to the client or not, should include the following:

1. Client name, date of estimate and name of the estimator.

2. All of the categories used in the estimating process. The number of categories and the delineation of specific costs is a function of how each individual business is run and specifically, whether items are subcontracted or whether labor and material are both purchased separately. Additionally, the source of particular items will determine whether those items are included in one particular category or another. As an example, the fireplace mantel may be included in millwork or it may be included in fireplaces, or it may be in a special category of its own, frequently called fireplace facings.

3. Provision for special items which will vary from job to job. These items will have no specific title other than "special" or "other", with space provided to insert a brief description of each item.

4. Specific provisions for miscellaneous and price escalation. There is nothing inappropriate about including a realistic allocation for each of these.

5. A rental equipment category, because certain jobs may require such items as concrete pumps or other specialized equipment.

ABC BUILDERS
123 Main Street
Hometown, MD 21234
(301) 231-1234

Client: _____

Date: _____

By: _____

GENERAL
architecture   _____

permits/bonds   _____

insurance   _____

direct supervision   _____

temporary utilities   _____

sani-john   _____

utility connections   _____

SITEWORK
survey/layout   _____

tree removal/clear   _____

septic/sewer - Allowance   _____

well/water connection - Allowance   _____

excavation/grade/backfill   _____

termite proof/demolition   _____

seed/landscape - Allowance   _____

FOUNDATIONS
footings   _____

sand/gravel   _____

structural steel   _____

waterproof/drain tile   _____

erosion control   _____

FRAMING
rough carpentry labor   _____

trim carpentry labor   _____

hardware labor   _____

rough lumber   _____

trim lumber   _____

stairs/rail   _____

WINDOWS/DOORS
interior doors   _____

exterior doors   _____

windows/fixed glass   _____

ROOFING
roofing

cornice

gutters/downspouts

PLUMBING
plumbing

HVAC
heat/air conditioning

humidifiers/air cleaners

whole house fans/attic fans

electrical

light fixtures - Allowance

INTERIOR FINISH
insulation

drywall

paint/stain

EXTERIOR FINISH
siding

brick siding

CONCRETE & MASONRY
concrete work

construction driveway

finish driveway - Allowance

culvert pipe/apron

garage doors

masonry

masonry wash

fireplaces

fireplace faces

INTERIOR FIXTURES
hardware/accessories - Allowance

cabinets/countertops - Allowance

vanity/mirrors/shower doors - Allowance

built-ins

appliances - Allowance

FLOORING
carpet – Allowance

hardwood

vinyl - Allowance

ceramic tile - Allowance

MISCELLANEOUS
decks - Allowance

security/alarms - Allowance

porch

intercom

central vacuum

trash pick-up

final clean

service work

rental equipment

theft, breakage, escalation, misc.

specials/other _____

specials/other _____

specials/other _____

TOTAL CONSTRUCTION COST
OVERHEAD/PROFIT

TOTAL COST

_____

Estimating Assumptions, Recommendations, and Bid Analysis

There is no better way of marketing yourself and your professionalism than in the presentation of the Bid Package to the client. In addition to the Summary Estimating Sheet previously discussed, every good estimate presentation should include Estimating Assumptions and Recommendations as well as a Bid Summary Analysis listing inclusions and exclusions. The Assumptions and Recommendations allow you to do two things:

1. All estimates require certain assumptions. By itemizing these assumptions for the client, you help avoid misunderstandings regarding the scope of the estimate.

2. The Recommendations allows you to point out to the client exactly how detailed a review you are capable of making and what an expert you are on residential construction.

It is important to study the plans carefully and find several areas with either minor errors or areas which provide an opportunity to make recommendations to improve the quality of the design. As the following examples indicate, very often people considering a custom home do not have a true idea of actual size and the scale represented by the drawings.

BUSINESS BUILDER:

There will never be a better opportunity to show your knowledge and design expertise than in the assumptions and recommendations attached to your estimate. Maximize this opportunity and show the client how much you know.

ABC BUILDERS
123 Main Street
Hometown, MD 21234
(301) 231-1234

**Estimating Assumptions/
Recommendations**

SAMPLE

Name of Residence

The following assumptions are noted:

1. Plans do not show walk out basement – per our meeting price quoted assumes one (1) 6' x 6' 8" Andersen sliding glass door and the grade marked in red on the returned plans and labeled "proposed grade".

2. Pump up basement plumbing assumed based on topography provided.

3. Plans show brick "water table" – assume standard brick as rowlock, no special shapes included in price.

4. Several elevations show selected windows without grills. Assume grills in all windows and doors.

5. Windows are labeled "Low E" glass. Assume all doors also to be "Low E" glass.

6. Plans show #5 rebar in garage slab. Our standard is #4. Price assumes #4 rebar. If #5 desired, add $120.00.

7. Plans do not show filled cores on block walls. It appears that grade may require one wall to hold more than 7 ft. of unbalanced fill. Price assumes no filled cores. If required, core fill is approximately $7.00/linear ft. of wall filled.

The following recommendations are noted:

1. Front elevation and 1st floor plan are in conflict – Labeled as C25 on floor plan, shown 6' higher elevation. Six foot assumed, modify floor plans to reflect this height.

2. Plan shows 42" vanity in bath #3. Recommend reducing to 36" vanity, allow more space at toilet (currently only 32").

3. Recommend reanalysis of kitchen. You should consider changing "U" shape to broken "L" with island – see attached sketch.

ABC BUILDERS
123 Main Street
Hometown, MD 21234
(301) 231-1234

SAMPLE

INCLUSIONS

* Brick to grade

* STK quality cedar siding

* White Andersen Low E windows
  with screen & grills

* Tile/marble per allowance

* Kohler Watersilk whirlpool
  model 1384 in standard color

* Concrete walk from front door
  to garage pad

* Master walk-in closet completely
  customized with unlimited
  shelves/poles

EXCLUSIONS

* Clear cedar siding

* Brick walkways

* Any dirt in or out of
  the site

* Retaining walls shown on
  site plan - see separate
  estimate

* "Future pool"

* Drawers in closets
  additional shelves/poles
  in other closets

Professional Services Agreement

A Professional Services Agreement provides an opportunity to protect yourself while you are doing preliminary work, and assures that you will be paid for your time. Also, we recommend strongly that you offer all the work included in the Professional Services Agreement as either a portion of your overhead and profit if you build the house or as a line item in the estimate. This makes your clients less anxious, since they know that if they proceed with the house they will not have spent additional monies. Similarly, it allows you to do additional professional work, safe in the knowledge that if they do not build the house, you will be paid under the terms of the agreement.

Every Professional Services Agreement should have the following:

1. A description of the services which you will provide a prospective client at no charge.

2. A description of services that are free of charge if you build the house for the client.

3. Any outside services which you will be contracting for in the early phases of the design. Such items may include engineering, architectural and survey work. This work typically, while paid for now, would show up as a line item on a detailed cost estimate. Thus, should the clients proceed, they will be credited with these monies against the original price estimate.

The use of a Professional Services Agreement also helps discourage those who are not serious and assures that you will not waste your time with people who have no real chance of becoming your clients. Anyone who is unwilling to put down an initial services payment of $500 to $1,000, knowing that this money will come back to them if they do build the house, is probably not a serious buyer.

PROFESSIONAL SERVICES AGREEMENT
Between ABC BUILDERS and _____

Dated: _____

ABC BUILDERS is a full-service custom home builder. We specialize in working with our clients in the early stages of the homebuilding process. The fee schedule for our professional services is divided into three categories as follows:

In-House Services at No Charge
- Initial meeting with client
- Computerized preliminary cost estimate
- Field inspection of lot with client to determine suitability for building

In-House Services Offered by ABC BUILDERS at No Additional Charge to ABC Clients Who Ultimately Sign a Contract with ABC BUILDERS to Construct Their Home
- Additional design consultations such as bathroom, kitchen, and lighting layout
- Engineering and structural analysis of plans
- Preparation of site plans
- Preparing permit applications for well, septic and building permits
- Revised cost estimates
- Preparation of draft contract and specifications

If the client does not sign a contract with ABC BUILDERS, these in-house services will be billed to the client at the rate of $60/hour.

Outside Architectural and Engineering Services to Be Paid by the Client
- Architectural renderings
- Preparation or modification of plans
- Survey work

The amount to be paid for any outside services shall be agreed upon in writing prior to those services being rendered.

AGREEMENT:

I hereby agree to pay ABC BUILDERS for in-house services rendered at the rate of $60/hour not to exceed the total sum of _____. I understand that if I sign a contract to construct a new home with ABC BUILDERS there will be no charge for these in-house services.

_____   _____
Owner   Date

_____   _____
Owner   Date

_____   _____
ABC BUILDERS   Date

I hereby agree to pay ABC BUILDERS for Outside Architectural and Engineering Services as follows:

Description of Service   Cost

_____   _____
Owner   Date

_____   _____
Owner   Date

_____   _____
ABC BUILDERS   Date

NOTE: ALL DRAWINGS, DESIGNS, AND ARCHITECTURAL REVISIONS TO STOCK PLANS ARE THE JOINT PROPERTY OF THE CLIENT AND ABC BUILDERS AND EACH PARTY HEREBY ACKNOWLEDGES THE RIGHT OF THE OTHER TO UTILIZE THEM.

ABC BUILDERS   **Building Permit Worksheet**
123 Main Street
Hometown, MD 21234
(301) 231-1234

Plans

_____1/4" scale drawings
_____foundation/footing plans
_____cross section
_____roof plan with structural
     information
_____structural plans

Site Plan

_____well location
_____septic field
_____topography
_____driveway and parking pad
_____building restriction lines
_____easements and setbacks

Application

_____Lot information
_____Lot and block
_____Subdivision
_____Street address
_____Tax I.D. number
_____Zoning
_____Acreage or square footage

_____Owner information
Name

Current address
Telephone number

_____Special requirements
Number of bedrooms
Number of full and half baths
Square footage
Height
Energy calculations

_____Bonds and fees
Impact fee calculations
Bond requirements including forms or
letters of credit
Application fee calculations

Sediment and Erosion Control/Storm Water Management

Disturbed area
Impervious area-runoff calculation
Grading plan
Sediment and erosion control plan
Drainage pipe sizes -- including
driveway

Five Reasons to Get Bids

Why get bids from subcontractors? There are five good reasons.

REASON #1: ACCURATE ESTIMATE

As a custom builder, you are going to be called upon to prepare estimates for your potential clients. These estimates generally lead to fixed-cost contracts. Your bottom-line profit margin depends on the accuracy of your estimate. Some parts of the estimate will be done in-house. In addition, we recommend that you obtain bids from your key subcontractors. By locking in these bids, the accuracy of your total estimate will be greatly improved.

REASON #2: COMPETITIVE BIDS

When you are estimating a house, you may want to get competitive bids to ensure that you are getting good prices from your regular subcontractors.

Using the standard bid forms contained in this section will ensure that you can compare apples to apples when you receive competitive bids.

Competitive bids from your subcontractor will help you produce an estimate that is accurate and competitive with other builders bidding the same job. Competitive bids can also help you control costs and increase profits for work that you already have under contract.

REASON #3: MINIMIZE EXTRAS

Using standard bid forms will clarify what is to be included in the scope of work for each subcontractor. This will eliminate debates concerning extra charges for work performed by subcontractors. The bid forms also include requests to submit labor rates for any extra work that is performed. You don't want any surprises like the plumber billing you at rates that look more like those of a doctor than a plumber!

REASON #4: SIMPLIFY SUBCONTRACTOR SUPERVISION

The standard bid forms contain a tremendous amount of information about the house. The forms are a consistent written record of what is to be included in each subcontractor's scope of work. The forms also clarify what materials are to be provided by ABC Builders and what materials are to be provided by the subcontractor. Having this information base simplifies the task of supervising the subcontractor. For example, if you have taken the time to specify that the bid/contract for the plumber is to include two 50-gallon A.O. Smith high-efficiency electric water heaters, you shouldn't have to spend supervisor time answering questions about what water heaters to install in the house.

REASON #5: IDENTIFY PROBLEMS

When you send out plans, specifications and bid forms to your key subcontractors, you gain the advantage of having many other sets of eyes reviewing the house prior to locking into a final proposal or contract with your client. Your subcontractors may identify problems that you have overlooked and that will cut into your profit. For example, you may have assumed that a two-zone heating/AC system is required but find out through the bid process that a three-zone system is required. Each of your key subcontractors can provide valuable information to you during the bid process that will ensure an accurate estimate and guarantee good profits.

Four-Step Bid Process

Bidding Procedure:
Use the following set of forms to organize the bidding procedure from your end and to clarify the scope of the job for the subcontractor.

STEP #1: BID ORGANIZATION FORM
Use this form to record and summarize the bids received for a specific job. The first part of the form can be used to show which subcontractors were sent bid packages and the remainder of the form can be filled in during the course of construction.

STEP #2: BID COVER SHEET
Use this form to send out with each bid package. It contains information about whom the bid was sent to, job name and location, enclosures to the bid package, bid due date and description of work to be bid.

STEP #3: STANDARD CONDITIONS
Use this form to ensure that the subcontractors know the "rules of the game." The standard conditions include information on insurance, labor and equipment, warranties, job cleanup, utilities, billing/payment and bid price.

STEP #4: SAMPLE BID FORMS
Sample bid forms are included for five key subcontractors -- masonry, rough carpentry, plumbing, electrical and HVAC. Each bid form covers the scope of the work to be performed, materials, inspections, unit prices and special items.

BUSINESS BUILDER

When you decide to submit a competitive bid, invest the energy and time necessary to get complete, accurate apples vs. apples bids. Anything less is a waste of time.

ABC BUILDERS
123 Main Street
Hometown, MD 21234
(301) 231-1234

JOB NAME:_____

TRADE	NAME	PHONE	PRICE	NOTES-------

ABC BUILDERS
123 Main Street
Hometown, MD 21234
(301) 231-1234

SENT TO:

NAME_____

ADDRESS _____

TELEPHONE _____

TRADE _____

JOB DESCRIPTION:

NAME _____

LOCATION _____

START DATE _____

INFORMATION ENCLOSED:

BID FORM   __ Yes __ No

PLANS   __ Yes __ No

SPECIFICATIONS   __ Yes __ No

BIDS DUE TO ABC BUILDERS ON: _____

DESCRIPTION OF WORK TO BE BID:

_____
_____
_____
_____

ABC BUILDERS
123 Main Street
Hometown, MD 21234
(301) 231-1234

**Standard Subcontractor Conditions
for Bidding**

The following items and conditions shall be included in the Subcontractor's bid:

Insurance:

Workmen's Compensation Amount _____

Liability Amount _____

<u>Original</u> certificates of insurance to be on file in ABC Builders office prior to first payment on subcontract.

Labor and Equipment:

Subcontractor shall provide all labor, tools and equipment required to perform the required work. Any exceptions shall be clearly specified in writing in the subcontractor's proposal/contract.

Warranty:

Subcontractor shall warrant all materials and workmanship furnished or performed to be free of defects for a period of one year from date of occupancy not to exceed eighteen months from completion of subcontractor's work or shall, at his own expense, promptly replace, repair or correct any such workmanship. All manufacturers' warranties for material supplied by the subcontractor shall be assigned to the contractor or owner. All rebates or incentive awards from manufacturers or utility companies shall be assigned to the contractor or owner.

Job Cleanup:

Subcontractor shall be responsible for daily cleanup of job and lunch trash. Trash shall be deposited in containers or locations supplied or designated by the contractor.

Utilities and Facilities:

Contractor shall supply -
     Portable toilets
     Dumpsters
     Trash cans
Subcontractor shall supply -
     Power except as noted in subcontract
     Water except as noted in subcontract

Billing and Payment:

Bills received by the 25th shall be paid by the 10th of the following month. Bills received by the 10th will be paid by the 25th of the month.

Bid Price:

The bid price shall be guaranteed for sixty days from time of submittal. The price shall include all labor, materials, overhead and profit as specified in the bid documents. <u>No</u> <u>extras</u> will be paid without <u>written</u> authorization of changes by the contractor.

123 Main Street
Hometown, MD 21234
(301) 231-1234

Scope of Work:

Block - Given corner points and footings level to ± 1/2" lay block in accordance with the plans including: subgrade block, 12" and 8" block, horizontal reinforcing every 16", fill solid under beam pockets, brick ledge at proper elevation, cap block, lintels and anchor straps or bolts. The subcontractor shall be responsible for laying out all opening and beam pockets in accordance with the plans and for verifying all dimensions. Any discrepancies shall be immediately noted to the contractor. All finish work shall be level and plumb ± 1/4".

Parging - All subgrade block shall be parged with 1/2" thick portland cement troweled smooth. All corner parging shall be extended two (2) feet past basement walls.

Brick - Lay brick in accordance with the plans including: arches, lintels, pullouts, quoins, retaining walls, steps, porches and flatwork. Brick to be struck and washed.

Fireplaces - Build fireplaces in accordance with plans including fireboxes, flue liners, dampers, ash dumps, outside air kits and chimney caps. Build all hearths and inside face brick in accordance with plans.

Unit Prices:

For any extra work not shown on the plans, please include in your bid unit prices for the following labor only items:

12" block	$_____	each
8" block	$_____	each
brick	$_____	thousands

Materials:

Materials shall be supplied according to the following schedule.

	Supplied By	
Item	ABC Builders or	Subcontractor
Sand and mortar	_____	_____
Block and lintels	_____	_____
Horz reinforcing and straps	_____	_____
Water	_____	_____
Brick $_____ /thousand	_____	_____
Fireplace materials	_____	_____
Brick cleaning	_____	_____

Inspections:
The subcontractor shall obtain inspections for parging and fireplace throats.

Miscellaneous: _____
_____
_____
_____

Protection of Work:
All glass shall be covered with 6 mil. poly prior to laying brick. All sills and concrete shall be covered with sand and cleaned after completion of brick work.

Special Items:_____
_____
_____
_____

Bids:
Please submit your bid broken out in the following categories:

Block $_____
Brick $_____
Fireplaces $_____

123 Main Street
Hometown, MD 21234
(301) 231-1234

Scope of Work:

General - Unless otherwise noted, the subcontractor shall provide all labor, material and equipment required to frame the house in accordance with the plans and specifications.

Steel - The subcontractor shall assemble and set all structural steel including lolly columns, beams, and flitch plates. If a crane is required, the subcontractor shall hire, supervise and pay for the crane. Whenever possible, ABC Builders will keep beam weights under 600 lbs.

First and Second Floor Decks - The subcontractor shall install sill sealer and plate material to basement walls. Plate material on beams shall be bolted or fastened by Hilti gun nails or equivalent. Install joists and band boards. All openings shall be as per plans. Any discrepancies shall be immediately reported to ABC Builders. Install, glue and nail or staple subfloor.

Exterior Walls - Frame, erect and fireblock exterior walls including headers with proper jacks, wall sheathing, corner and wall backer details coordinated with ABC Builders. Tyvek or equivalent shall be applied using approved lap, corner wrapping and window opening techniques.

Interior Walls - Install all interior partitions including blocking for drywall with proper door openings.

Roof Framing - Set roof trusses and conventional framing in accordance with the plans. Apply plywood and "H" clips. Cut openings for skylights, ridge vent, attic fans, chimneys. Build all required crickets and watersheds. If a crane is required, the subcontractor shall hire, supervise and pay for the crane.

Blocking - Frame all access panels, chases, block for stair rails, towel bars, medicine chest, laundry chutes.

Boxing and Cornice - Box HVAC ducts, kitchen and bath bulkheads, install rake and gutter trim boards and box for freeze board.

Windows, Doors and Stairs - Set all exterior windows, doors, skylights and prepare framing for garage doors. Set interior stairs including building platforms and support walls.

Exterior Decks and Screen Porches

Unit Prices:

For any extra work please include in your bid unit prices which shall be used to bill for labor only work.

Lead carpenter and truck	$_____	/hr.
Carpenter's helper	$_____	/hr.
Laborer	$_____	/hr.

Materials:

Materials shall be supplied according to the following schedule:

	Supplied By	
Item	ABC Builders or	Subcontractor
Steel	_____	_____
Framing Lumber	_____	_____
Nails	_____	_____
Fasteners & Glue	_____	_____
Other	_____	_____
Other	_____	_____

Inspections:

The subcontractor shall be responsible for correcting any deficiencies found during the building inspector's framing inspection which are not related to errors in the plans and specifications.

Miscellaneous: _____

_____

_____

_____

123 Main Street
Hometown, MD 21234
(301) 231-1234

Scope of Work:
General - Unless otherwise noted, the subcontractor shall provide all labor,
material and equipment required to plumb the house in accordance with the plans and
specifications. The subcontractor shall obtain all permits and inspections.

Water and Sewer - Connect the house to public water and sewer.
Water Line: size _____ material _____ length _____

Sewer Line: size _____ material _____ length _____ depth _____

Well and Septic - Connect the house to well and septic tank including well pump, electric
line and ground wire if required.

Water Line - length including depth of well _____ ft.

Pump Size - _____ hp.

Septic Line - length to tank _____ ft.

Materials:
Supply Lines _____

Drain Lines _____

Vent Lines _____

Soil Lines _____

Supply and Install the following:
Equipment and Rough In

Hose Bibs _____   Laundry Trays _____   Washer Box _____
Water Heaters _____
Pressure Tank _____
Ejector Pit and Pump _____
Sump Pump _____   Basement Rough-In Full Bath _____
Vinyl Shower Pans _____

Plumbing Fixtures and Trim:

Description	Quantity	Brand	Model

Bath Sinks

_____

_____

Bath Sink Faucets

_____

_____

Toilets/Bidets

_____

_____

Tubs

_____

_____

Tub and Shower Faucets

_____

_____

Kitchen and Specialty Sinks

_____

_____

Install Only
Ice maker lines _____
Humidifier lines _____
Disposals _____
Dishwashers _____
Washing machine _____
Instant hots _____

Unit Prices - For any extra work not shown on the plans and specifications, please include in your bid unit prices for the following labor only items:
Plumber $_____/hr.
Plumber's helper $_____/hr.

Miscellaneous and Special Items: _____

_____

_____

123 Main Street
Hometown, MD 21234
(301) 231-1234

Scope of Work:

General - Unless otherwise noted, the subcontractor shall provide all labor, materials and equipment required to wire the house in accordance with the plans and specifications. The subcontractor shall obtain all permits and inspections.

All wiring shall be done in accordance with the applicable national and local electric codes. It shall be the subcontractor's responsibility to obtain all inspections and notify the contractor of any discrepancies in the plans and specifications that do not meet code requirements.

Rough Wire - The subcontractor shall rough wire the house including intercom, TV, telephone and speaker wires as required and set the panel.

Final Wire - The subcontractor shall final wire the house including installation of all fixtures and appliances and obtain all final inspections.

Unit Prices - For any extra work please include in your bid the following unit prices which shall be used to bill for labor only work.

Electrician and truck $_____/hr
Electrician's helper $_____/hr

The attached unit price sheet shall be completed and kept on file with ABC Builders. The subcontractor agrees to bill all extras in accordance with the unit price sheet.

Materials - The subcontractor shall supply all materials except as noted/clarified in Table A.

TABLE A

ELECTRICAL REQUIREMENTS
Service
200 amps _____ 300 amps _____ 400 amps _____ Other _____

Equipment Connections
Compressors _____
Air Handle(s) incl. resistant heat _____
Water Heater(s) _____
Air Cleaner _____ Humidifier _____ Central Vac _____
Alarm _____ Sump Pump _____ Ejector Pump _____
Attic Fan(s) _____ Whole House Fan _____ Baseboard Heater(s) _____
Other _____
_____

Appliances
Dishwasher _____ Disposal _____ Instant Hot _____
Range _____ Wall Oven _____ Microwave _____
Refrigerator _____ Freezer _____ Trash Compactor _____
JennAire _____ Range Hood _____ Electric Dryer _____
Gas Dryer _____ Washer _____ Pull-Down Iron _____
Other_____
_____

Lighting
Recessed _____ Surface Mount _____ Track _____
Fluorescent _____ Porcelain _____ Floodlights _____
Chandeliers _____ Under Counter _____
Exterior Wall Mount _____ Exterior Post/Column _____
Driveway Lights _____ Ft. of Trenching _____

Low Voltage
Intercom _____ Master _____ Speakers
_____Telephone
_____TV _____Speaker(s)

Fans
Bath Fans _____ Bath Fan Lights _____ Paddle Fans _____

Qty.	ITEM	PRICE
1	PERMIT	$ 0.00
1	RECEPTACLES	$13.50
1	SWITCHED RECEPTACLES	$17.00
1	GFI RECEPTACLES	$33.50
1	W.P. RFCEPTACLES	$18.50
1	ISLAND RECEPTACLES	$28.50
1	FLOOR RECEPTACLES	$43.50
1	1-POLE SWITCHES	$13.50
1	3-WAY SWITCHES	$17.00
1	4-WAY SWITCHES	$33.50
1	SURFACE MOUNT LIGHTS	$13.50
1	WALL MOUNTED OUTSIDE LIGHTS	$18.50
1	PULL CHAIN LIGHTS	$16.50
1	RECESSED LIGHTS	$19.50
1	TRACK LIGHTS	$43.50
1	UNDER-CABINET LIGHTS	$28.50
1	FLOODLIGHTS	$43.50
1	CATHEDRAL OUTLETS	$17.00
1	PADDLE FANS	$48.50
1	EXHAUST FANS	$13.50
1	FAN/LIGHT COMBINATIONS	$27.00
1	ATTIC VENT FANS	$18.50
1	SMOKE DETECTORS	$28.50
1	TELEPHONE OUTLETS	$18.50
1	WALL TELEPHONE OUTLETS	$19.50
1	CABLE TV OUTLETS	$30.00
1	WHIRLPOOL BATH	$125.00
1	12 KW RANGE	$85.00
1	COOKTOP	$75.00
1	WALL OVEN	$90.00
1	MICROWAVE OUTLET	$50.00
1	DISHWASHER	$50.00
1	GARBAGE DISPOSAL	$55.00
1	TRASH COMPACTOR OUTLET	$50.00
1	INSTA-HOT OUTLET	$50.00
1	WASHER OUTLET	$50.00
1	DRYER OUTLET	$75.00
1	WATER HEATER	$70.00
1	DOOR BELL	$45.00
1	WELL PUMP	$60.00
1	15-20KW HEAT PUMP SYSTEM	$310.00
1	10-15KW 2ND FL HVAC	$350.00

1	25KW HEAT PUMP SYSTEM	$420.00
1	200 AMP SERVICE	$310.00
1	300 AMP SERVICE	$515.00
1	400 AMP SERVICE	$720.00
1	INSTALL METER	$85.00
1	600 WATT DIMMER	$10.00
1	1000 WATT DIMMER	$55.00
1	BOND WHIRLPOOL BATHS	$35.00
1	VACUUM CIRCUIT	$45.00
1	ATTIC & CRAWL SPACE LIGHT	$13.50

123 Main Street
Hometown, MD 21234
(301) 231-1234

Scope of Work:

General - Unless otherwise noted, the subcontractor shall provide all labor, materials and equipment required to heat, air condition and ventilate in accordance with the plans and specifications.

Type of Equipment - Options Include:

Heat Pump with Electrical Resistance Backup	HP/Elec
Heat Pump with Oil Backup	HP/Oil
Heat Pump with Gas Backup	HP/Gas
Gas with Air Conditioning	Gas/AC
High-Efficiency Gas with Air Conditioning	HEGas/AC
Oil with Air Condition	Oil/AC
Forced Air Elec. Resistance with Air Conditioning	Elec/AC
Baseboard Elec. with Air Conditioning	BBElec/AC
Other_____	
Other_____	

For this bid the type of equipment is as follows:

	Location	Type
Zone 1	_____	_____
Zone 2	_____	_____
Zone 3	_____	_____
Zone 4	_____	_____

Size of Equipment - The bid shall specify the size and manufacturer for each zone of the house.

Ductwork - The bid shall specify the number of central returns, individual returns and supplies for each system.

Flues - Flues for oil and gas shall be as follows:
Separate flue _____
Tie into masonry flue _____
High efficiency through the wall or roof _____

Ventilation - The bid shall include the following ventilation:

Bath Fans _____ Model(s) _____

Bath Fans/Lights _____ Model(s) _____

Bath Heat/Fan Lights _____ Model(s) _____

Dryer Vent _____

JennAir Vent _____

Wall Oven Vent _____

Other _____

_____

Other Equipment - The bid shall include the following:

Humidifiers _____ Model(s) _____

Electronic Air Cleaners _____ Model(s) _____

Night Set-Back Thermostat _____ Model(s) _____

Oil Tank _____

Unit Prices - For any extra work not shown on the plans and specifications, please include in your bid unit prices for the following labor only items:

Lead Installer $_____ /hr.

Helper $_____ /hr.

Miscellaneous and Special Items - _____

_____

_____

_____

CONSTRUCTION PACKAGE

Construction Schedule Form

This is a simple scheduling system based on a Weekly Plan. At the beginning of the week you indicate the tasks to be accomplished in the first column, and then decide which tasks will be completed on which days in the other columns. The tasks are based on a 22-week or a 30-week construction schedule. Use whichever schedule is appropriate. Each task is broken down by week. You can modify this schedule to better suit the way you build.

Rather than rewrite all the tasks, you can cut a copy of the schedule apart week by week, and then paste it in the first column, then use the task numbers to indicate which tasks will be completed on the different days. At the bottom of the column is a space for a daily log. If your comments exceed the space allotted, make them on a second piece of paper and attached it to the form.

Each task can be dealt with in several ways. You can complete it, and mark it done. You can transfer it to another date (indicate the date), you can decide it doesn't need to be done and cancel it, or you can delegate it to someone on the staff. If you delegate it, indicate the person to whom it has been delegated. The symbols for each of these ways of dealing with a task are indicated at the top of the page.

All tasks are also marked in terms of priority. "A" tasks are tasks that should be completed that day. They are also tasks that will have a major impact on the job schedule or on profits. "B" tasks can often be deferred to the next day. They are less critical in terms of schedule or profits. "C" tasks can be put off for several days, if need be. Always work on all "A" tasks before working on "B" tasks, etc.

Make copies of the weekly planning form and put them in a 3-hole binder. This becomes your weekly planning tool. At the end of the week, make a copy of the form along with your comments, and send it back to the main office as a schedule report form.

NOTE: When printing the schedule form below it is probably necessary to change settings for the printer. The form needs to be printed in landscape view.

Month _____ Builder Weekly Plan _ Done ➔ Transferred ✗ Cancelled O Delegated   PRIORITY: A B C then number "A" tasks.

Week _____	Monday_____	Tuesday _____	Wednesday_____	Thursday _____	Friday _____	Saturday_____
Weekly Plan	To Do	To Do	To Do	To Do	To Do	To Do
						Daily Log
	Daily Log	Daily Log	Daily Log	Daily Log	Daily Log	Sun

ABC BUILDERS
123 Main St.
Hometown, MD 21234
(301) 231-1234

PRE-CONSTRUCTION
0.1 Meet new clients
0.2 Prepare or redraw plans
0.3 Prepare estimate
0.4 Prepare draft contract/specifications
0.5 Review drawings
0.6 Prepare final cost estimate/contact and specs
0.7 Apply for electrical service
0.8 Submit building permit including site plan with septic information
0.9 Prepare to drill well and stake well location
0.10 Drill well

WEEK 1
1.1 House stakeout
1.2 Phone for clearing
1.3 Flag for clearing
1.4 Clear lot

WEEK 2
2.1 Order basement windows/doors
2.2 Phone for foundation excavation
2.3 Obtain building permit
2.4 Excavation layout
2.5 Phone for footings
2.6 Excavate foundation
2.7 Phone mason for block delivery
2.8 Deliver basement windows/doors

WEEK 3
3.1 Footings
3.2 Phone for roof trusses
3.3 Phone for footing inspection
3.4 Phone for structural steel/drain tile/columns
3.5 Nail points in footings
3.6 Phone for underslab plumbing/laterals
3.7 Phone for concrete slab work

WEEK 4
4.1 Deliver block
4.2 Start foundation walls
4.3 Phone for partial backfill
4.4 Phone carpenter
4.5 Phone for wall check survey
4.6 Order basement gravel
4.7 Complete foundation walls
4.8 Phone for 1st floor lumber package
4.9 Deliver structural steel
4.10 Parge foundation walls

WEEK 5
5.1 Underslab plumbing
5.2 Wall check survey
5.3 Phone for parging inspection
5.4 Underslab plumbing inspection
5.5 Partial backfill

WEEK 6
6.1 Grade basement subgrade and gravel
6.2 Pour basement floor
6.3 Inspection for slab
6.4 Grade garage subgrade
6.5 Dig and install W&S laterals or well and septic
6.6 Deliver 1st floor lumber package
6.7 Set structural steel
6.8 Pour garage slab

WEEK 7
7.1 Frame 1st floor deck
7.2 Grade crawl space
7.3 Phone for 2nd floor lumber package/roof lumber
7.4 Phone for windows and exterior doors
7.5 Frame 1st floor walls

WEEK 8
8.1 Deliver 2nd floor lumber
8.2 Deliver 2nd floor lumber package/roof lumber
8.3 Frame 2nd floor deck
8.4 Phone for stair/rail measurement
8.5 Frame 2nd floor walls
8.6 Phone for brick veneer

WEEK 9
9.1 Frame garage
9.2 Deliver roof trusses
9.3 Phone utility re URD cable
9.4 Set roof trusses
9.5 Order shingles

WEEK 10
10.1 Sheeth roof
10.2 Phone for plumbing rough-in
10.3 Phone for backfill and rough grade
10.4 Deliver windows and exterior doors
10.5 Phone heating and A/C rough-in

WEEK 11
11.1 Set windows and exterior doors
11.2 Deliver stairs
11.3 Backfill and rough-grade
11.4 Phone roofers
11.5 Phone for exterior trim package
11.6 Phone for exterior siding material
11.7 Begin brick veneer
11.8 Set interior stairs
11.9 Phone for siding installers
11.10 Roofing

WEEK 12
12.1 Order hardware package
12.2 Plumbing rough-in
12.3 Phone for electric rough-in
12.4 Order hardwood flooring

WEEK 13
13.1 Prepare interior finish schedule/door order
13.2 Order kitchen cabinets and A/C rough-in
13.3 Heating and A/C rough in

WEEK 14
14.1 Interior framing backup incl. soffits/box ducts
14.2 Phone for garage door
14.3 Phone for exterior trim
14.4 Deliver exterior trim package
14.5 Electric rough-in
14.6 Begin exterior trim (cornice)
14.7 Phone for concrete steps
14.8 Phone for insulation
14.9 Phone for drywall delivery and installation
14.10 Install water meter
14.11 Install furnace
14.12 Complete exterior trim (cornice)
14.13 Obtain electrical/plumbing rough-in inspections

WEEK 15
15.1 Call for framing inspection
15.2 Framing inspection
15.3 Hang garage doors
15.4 Form concrete steps
15.5 ABC framing inspection
15.6 Frame punch-out
15.7 Pour concrete steps
15.8 Install insulation
15.9 Call for insulation inspection
15.10 Order light fixtures
15.11 Phone for interior door and trim package
15.12 Insulation inspection
15.13 Deliver drywall
15.14 Install electric meter
15.15 Phone for gutters and downspouts
15.16 Phone for drywall hanging

WEEK 16
16.1 Turn on heat
16.2 Hang and tape drywall
16.3 Phone for driveway
16.4 Phone for exterior painting
16.5 Order ceramic tile
16.6 Hang gutters and downspouts
16.7 Exterior painting
16.8 Phone kitchen cabinet/vanities delivery

WEEK 17
17.1 1st coat drywall
17.2 Phone for final grading
17.3 Phone for rail delivery
17.4 Order kitchen appliances
17.5 Last date for paint selection
17.6 Deliver wood flooring
17.7 Lay wood floors

WEEK 18
18.1 Deliver interior doors and interior trim
18.2 Interior trim
18.3 Final grade by excavator
18.4 Deliver kitchen cabinets & vanities
18.1 Install kitchen cabinets and vanities
18.2 Measure/order countertops
18.3 Phone for seeding or sodding
18.5 Install ceramic tile
18.6 Phone for switch and plug
18.7 Phone interior paint/caulk
18.8 Deliver hardware package

WEEK 19
19.1 Prime paint
19.2 Drywall point-up/sand
19.3 Fine grade yard for seed
19.4 Phone for final HVAC
19.5 Phone floor finishing
19.6 Deliver countertops
19.7 Set countertops
19.8 Deliver kitchen appliances
19.9 Sand and finish floors

WEEK 20
20.1 Phone final survey
20.2 Paper hardwood floors
20.3 Deliver electrical fixtures
20.4 Hang electrical fixtures
20.5 Final HVAC startup/grills
20.6 Set plumbing fixtures
20.7 Set kitchen appliances

WEEK 21

21.1 Call for water test
21.2 Lay carpet/vinyl
21.3 Call for final clean
21.4 Hang electrical fixtures
21.5 Final survey
21.6 Phone for A/C or heat check
21.7 Set finish hardware
21.8 Hang mirrors

WEEK 22

22.1 Hang window screens/grills
22.2 Phone for final clean
22.3 Install shoe mold
22.4 Final clean
22.5 Walk-thru with client
22.6 Call for final inspection
22.7 Final inspection
22.8 Issue final occupancy permit

ABC BUILDERS
123 Main St.
Hometown, MD 21234
(301) 231-1234

PRE-CONSTRUCTION
0.1 Meet new clients
0.2 Prepare or redraw plans
0.3 Prepare estimate
0.4 Prepare draft contract/specifications
0.5 Review drawings
0.6 Prepare final cost estimate/contact and specs
0.7 Apply for electrical service
0.8 Submit building permit including site plan with
   septic information
0.9 Prepare to drill well and stake well location
0.10 Drill well

WEEK 1
1.1 House stakeout
1.2 Phone for clearing
1.3 Flag for clearing
1.4 Clear lot

WEEK 2
2.1 Order basement windows/doors
2.2 Phone for foundation excavation
2.3 Obtain building permit
2.4 Excavation layout
2.5 Phone for footings
2.6 Excavate foundation
2.7 Phone mason for block delivery
2.8 Deliver basement windows/doors

WEEK 3
3.1 Footings
3.2 Phone for roof trusses
3.3 Phone for footing inspection
3.4 Phone for structural steel/drain tile/columns
3.5 Nail points in footings
3.6 Phone for underslab plumbing/laterals
3.7 Phone for concrete slab work

WEEK 4
4.1 Deliver block
4.2 Start foundation walls
4.3 Phone for partial backfill
4.4 Phone carpenter
4.5 Phone for wall check survey
4.6 Order basement gravel

WEEK 5
5.1 Complete foundation walls
5.2 Phone for 1st floor lumber package
5.3 Deliver structural steel
5.4 Parge foundation walls

WEEK 6
6.1 Underslab plumbing
6.2 Wall check survey
6.3 Phone for parging inspection
6.4 Underslab plumbing inspection
6.5 Partial backfill

WEEK 7
7.1 Grade basement subgrade and gravel
7.2 Pour basement floor
7.3 Inspection for slab
7.4 Grade garage subgrade

WEEK 8
8.1 Dig and install W&S laterals or well and septic
8.2 Deliver 1st floor lumber package
8.3 Set structural steel
8.4 Pour garage slab

WEEK 9
9.1 Frame 1st floor deck
9.2 Grade crawl space
9.3 Phone for 2nd floor lumber package/roof lumber

WEEK 10
10.1 Phone for windows and exterior doors
10.2 Frame 1st floor walls

WEEK 11
11.1 Deliver 2nd floor lumber
11.2 Deliver 2nd floor lumber package/roof lumber
11.3 Frame 2nd floor deck
11.4 Phone for stair/rail measurement

WEEK 12
12.1 Frame 2nd floor walls
12.2 Phone for brick veneer

WEEK 13
13.1 Frame garage
13.2 Deliver roof trusses
13.3 Phone Utility re: URD cable
13.4 Set roof trusses
13.5 Order shingles

WEEK 14
14.1 Sheeth roof
14.2 Phone for plumbing rough-in
14.3 Phone for backfill and rough grade
14.4 Deliver windows and exterior doors
14.5 Phone heating and A/C rough-in

WEEK 15
15.1 Set windows and exterior doors
15.2 Deliver stairs
15.3 Backfill and rough-grade
15.4 Phone roofers
15.5 Phone for exterior trim package
15.6 Phone for exterior siding material

WEEK 16
16.1 Begin brick veneer
16.2 Set interior stairs
16.3 Phone for siding installers
16.4 Roofing

WEEK 17
17.1 Order hardware package
17.2 Plumbing rough-in
17.3 Phone for electric rough-in
17.4 Order hardwood flooring

WEEK 18
18.1 Prepare interior finish schedule/door order
18.2 Order kitchen cabinets and A/C rough-in
18.3 Heating and A/C rough in

WEEK 19
19.1 Interior framing backup inc. soffits/box ducts
19.2 Phone for garage door
19.3 Phone for exterior trim
19.4 Deliver exterior trim package
19.5 Electric rough-in
19.6 Begin exterior trim (cornice)
19.7 Phone for concrete steps
19.8 Phone for insulation
19.9 Phone for drywall delivery and installation
19.10 Install water meter
19.11 Install furnace
19.12 Complete exterior trim (cornice)
19.13 Obtain electrical/plumbing rough-in

WEEK 20
20.1 Call for framing inspection
20.2 Framing inspection
20.3 Hang garage doors
20.4 Form concrete steps
20.5 ABC framing inspection
20.6 Frame punch-out
20.7 Pour concrete steps

WEEK 21
21.1 Install insulation
21.2 Call for insulation inspection
21.3 Order light fixtures
21.4 Phone for interior door and trim package
21.5 Insulation inspection
21.6 Deliver Drywall
21.7 Install electric meter
21.8 Phone for gutters and downspouts
21.9 Phone for drywall hanging

WEEK 22
22.1 Turn on heat
22.2 Hang and tape drywall
22.3 Phone for driveway
22.4 Phone for exterior painting
22.5 Order ceramic tile
22.6 Hang gutters and downspouts
22.7 Exterior painting
22.8 Phone kitchen cabinet/vanities delivery

WEEK 23
23.1 1st coat drywall
23.2 Phone for final grading
23.3 Phone for rail delivery
23.4 Order kitchen appliances
23.5 Last date for paint selection
23.6 Deliver wood flooring
23.7 Lay wood floors

WEEK 24
24.1 Deliver interior doors and interior trim
24.2 Interior trim
24.3 Final grade by excavator
24.4 Deliver kitchen cabinets & vanities

WEEK 25
25.1 Install kitchen cabinets and vanities
25.2 Measure/order countertops
25.3 Phone for seeding or sodding
25.5 Install ceramic tile
25.6 Phone for switch and plug
25.7 Phone interior paint/caulk
25.8 Deliver hardware package

WEEK 26
26.1 Prime paint
26.2 Drywall point-up/sand
26.3 Fine grade yard for seed
26.4 Phone for final HVAC
26.5 Phone floor finishing

WEEK 27
27.1 Deliver countertops
27.2 Set countertops
27.3 Deliver kitchen appliances
27.4 Sand and finish floors
27.5 Phone final survey
27.5 Paper hardwood floors

WEEK 28
28.1 Deliver electrical fixtures
28.2 Hang electrical fixtures
28.3 Final HVAC startup/grills
28.4 Set plumbing fixtures
28.5 Set Kitchen appliances

WEEK 29
29.1 Call for water test
29.2 Lay carpet/vinyl
29.3 Call for final clean
29.4 Hang electrical fixtures
29.5 Final survey
29.6 Phone for A/C or heat check
29.7 Set finish hardware
29.8 Hang mirrors

WEEK 30
30.1 Hang window screens/grills
30.2 Phone for final clean
30.3 Install shoe mold
30.4 Final clean
30.5 Walk-thru with client
30.6 Call for final inspection
30.7 Final inspection
30.8 Issue final occupancy permit

Subcontracts

The documents used between builders and subcontractors are frequently provided by the subcontractor. In this case, the documents should be carefully reviewed and modified to protect the builder. A better way to do business is to prepare standard contracts for your major subcontractors and tell all new prospective subcontractors that this is the form you use. Such agreements should include the following:

1. **Spell out exactly what you will do**; for example, provide temporary heat or power or trash containers.

2. **The responsibilities of the subcontractor**. This should include specifically whether or not he is providing labor, equipment or both, the specifications to which he must adhere, a brief description of the scope of his work and his obligations for keeping the job clean.

3. **Any items which have historically been items of debate or discussion**. Delineating it here ends such debate about which trade is responsible for getting it done.

4. **When the subcontractor will start and approximately how long it will take to accomplish the work**.

5. **A payment schedule including draw amounts and the specific work required to be completed for a draw**.

The subcontract should have provisions for signatures and dates of both parties and copies should be provided to both the subcontractor and the builder. Certainly, many small builders use the same subcontractors over and over again and there is a great temptation to do this on a verbal contract or on a form simpler than the one shown. The good builder must resist this temptation, because regardless of the relationship between the builder and subcontractor, the reality is that a handshake is good enough when everything is going well, but a contract is what you need when a true dispute arises.

ABC BUILDERS
123 Main Street
Hometown, MD 21234
(301) 231-1234

Trim Carpentry Subcontract

Subcontract between _____, hereinafter referred to as "Trim Carpenter" and ABC BUILDERS, hereinafter referred to as "Contractor," dated_____, for work to be performed on the _____ Residence located at _____.

The Contractor shall be responsible for the following items:
- Providing Plans, Specifications and Order List and Trim Schedule.
- Supplying all materials except nails.
- Providing Power and Temporary Heat as required.
- Providing scaffolding and trash containers.

The Trim Carpenter shall be responsible for the following items:
- Providing all labor and equipment required to do the trim carpentry work in accordance with the Plans and Specifications and installing all materials shown on the Trim Order List.
- Setting all kitchen cabinets, kitchen countertops, vanities, vanity tops, desks and bars in accordance with the Kitchen Plans and Specifications.
- Putting all trash in containers and placing containers in garage.
- Leaving the house broom clean at completion of each work week.
- Building all exterior decks in accordance with the Plans and Specifications.

The Trim Carpenter shall start work within three (3) working days after being notified that the house is ready for trim and the materials are delivered to the site. The work shall be completed within approximately _____ working days. The Trim Carpenter shall be paid the following amounts for labor and equipment only:

Interior Trim	$_____	Lump Sum.
Kitchen and Vanities	$_____	Lump Sum.
Decks	$_____	Lump Sum.

The Trim Carpenter shall be paid within three (3) days after completion of the following items on the Trim Draw Schedule:

Items Needing to be Completed	Amount to be Paid
Doors, Windows, Closets, Base	60% of Interior Trim (see above)
Underlayment, Handrails, all trim	20% of Interior Trim (see above)
Mirrors, Medicine Chests, Locks, etc.	20% of Interior Trim (see above)
Cabinets set	80% of Kitchens & Vanities (see above)
Counter tops set	20% of Kitchens & Vanities (see above)
Deck(s) - when complete	100% of Decks (see above)

The above prices, specifications and conditions are satisfactory and are hereby accepted. The Trim Carpenter is authorized to do the work as specified. Payment will be made as outlined above.

_____   _____
ABC Builders   Date
Signed by: _____

_____   _____
Trim Carpentry Firm   Date
Signed by: _____

ABC Builders
123 Main Street
Hometown, MD 21234
(301) 231-1234

Frame Carpentry Subcontract

Subcontract between _____, hereinafter referred to as "Frame Carpenter" and ABC BUILDERS hereinafter referred to as "Contractor", dated _____, for work to be performed on the _____ Residence located at _____.

The Frame Carpenter shall furnish the materials and perform the labor necessary for the completion of the following items:

- Framing house in accordance with the plans.
- Setting steel beams, columns.
- Framing floors.
- Erecting interior and exterior walls.
- Framing roof.
- Setting roof trusses.
- Sheathing roof.
- Setting exterior windows, doors and interior stairs.
- Blocking for access panels, fire stopping, chases, stair rails, towel bars, medicine chests.
- Boxing HVAC ducts and kitchen soffit, rake, gutter and facia trim boards.
- Leaving work broom clean with all scrap material stacked in one location designated by the Contractor.
- Providing generator, if required.

The Contractor shall provide all material and supervision required to complete the work described above, not including nails or generator.

Extras will be charged at the rate of $_____ per man-hour.

All material is guaranteed to be as specified, and the above work to be performed in accordance with the drawings and specifications submitted for above work and completed in a substantial workmanlike manner for the sum of _____ _____ Dollars ($_____).

Payments to be made as follows:

Any alteration or deviation from above specifications involving extra costs will be executed only upon written orders, and will become an extra charge over and above the estimate. All agreements contingent upon strikes, accidents or delays beyond the control of the Frame Carpenter.

Owner to carry fire, tornado and other necessary insurance upon above work. Workmen's Compensation and Public Liability Insurance on above work to be taken out by _____. The above prices, specifications and conditions are satisfactory and are hereby accepted. The Frame Carpenter is authorized to do the work as specified. Payment will be made as outlined above.

_____   _____
ABC Builders   Date
Signed by: _____

_____   _____
Frame Carpentry Firm   Date
Signed by: _____

Draw Schedules

Various types of draw schedules are used in the construction industry. The two most common types are line item draw schedules and specifically described phased draw schedules.

In the line item draw schedule, the project will be divided into many small items, each of which will be assigned a dollar value or percentage of the contract amount. There may be as few as ten or twelve items to as many as thirty or forty. Typically, a list such as this will include approximately twenty-five to thirty items. Each will have a percentage equivalent of between 1% and 10% of the total job cost. In this methodology, the builder asks for a draw and specifies which line items are complete. This method frequently allows for a non-fixed number of draw requests, although some institutions will have a minimum allowable draw (dollar value or percent). For example, the builder may not ask for a draw of less than 10%.

In the second type of draw schedule, the draw will be divided into segments, and each segment will have a description of the items which must be done to receive the payment for that segment. Each segment is then assigned a percentage or dollar value and a draw is requested when all the items in that segment are complete. More liberal banks will often allow trade-offs between segments. For example, if a segment had everything completed except several doors were missing, but if half brickwork had been done from the next draw, the bank may trade the brick for the missing doors, and, thus, distribute the draw that included the doors as a small portion of the total list of items in that phase. Draw schedules of this type frequently divide the project into five to nine parts, with the most common draw schedules being five, six or seven phases.

All draw schedules should have the following characteristics:

1. The job should be divided into a reasonable number of parts so that the builder does not have to go a long time without getting funding for the job.

2. The items should be broken out in such a way that they mesh with the builder's methods of construction. It does no good to have a requirement to have a basement poured in the first draw of a phase draw schedule if, typically, the builder does not pour the basement until after the roof is on the house.

In many areas of the country, the builder does not have any input into the draw schedule and has to live with whatever the local lender wishes to enforce. Other builders are more fortunate -- lenders will work with them and use a schedule which is acceptable to all parties. This should always be your goal, and with that draw schedule you should try to create a document which will allow you to get money in reasonable quantities at reasonable intervals.

BUSINESS BUILDER

Remember, a draw schedule is designed as a tool to ensure that you are paid as the work progresses. It is neither a religious implement to be worshipped nor a whip with which the lender may punish the builder. It is a tool designed to help all parties, and it is a wise builder who refuses to sign an unfair draw schedule.

ABC Builders
123 Main Street
Hometown, MD 21234
(301) 231-1234

Construction Draw Schedule

Client _____

Contract Amount _____

Less Deposit _____

Draw #	Description of Work	Percent of Draw	Amount
1	When the first floor deck or slab is complete, no basement, foundation survey received, building permit issued and noted, proof of satisfactory Builder's Risk Insurance Policy	15%	_____
2	When second floor deck is complete or roof trusses, as the case may be	10%	_____
3	When house is under roof with roofing applied, windows and exterior door frames are in place	15%	_____
4	When plumbing, heating and electrical systems have been roughed in, and all siding and/or brickwork has been completed	20%	_____
5	When drywall has been applied and basement floors have been poured	10%	_____
6	When house is completely trimmed out, HVAC complete, kitchen cabinets and ceramic tile complete and all grading and cement work have been completed	20%	_____
7	When house is substantially complete with Use and Occupancy Permit, affidavit that all bills for material and labor have been properly listed and all liens have been waived on payment of said bills	10%	_____
TOTAL		100%	_____

ACCEPTED AND AGREED:

_____   _____
Owner   Date

_____   _____
Owner   Date

_____   _____
ABC BUILDERS   Date

Functions

To truly understand the operation and functions of any company or organization, it is necessary to think through the entire process of the day-to-day running of that company. All of the tasks necessary for accomplishing the goals of the organization must be listed. In a custom building business, this encompasses everything from meeting new clients through customer service work at the completion of the home.

The attached list is an example of an attempt to list all of the functions for a particular company with emphasis on the construction process. <u>This list is superintendent oriented.</u> A list such as this can be constructed, in detail, for all employees, from president to secretary. To create such a list for your own company, it is usually helpful to arrange day-to-day activities in the basic sequence in which they take place, from the beginning -- meeting the client -- to the end -- service after move-in.

In addition, it is often helpful to take this function list and use it to determine who in the organization is responsible for which function. This can be done through color coding the different functions. Another method is to assign numbers to each task, with each number representing a particular person in the organization. This will often help evolve clear lines of responsibility and avoid duplication of effort.

ABC BUILDERS
123 Main Street
Hometown, MD 21234
(301) 231-1234

----- Meet new clients
----- Prepare or redraw plans
----- Prepare estimate
----- Prepare draft contract/specifications
----- Review drawings
----- Prepare final cost estimate/contract and specs
----- Apply for electrical service
----- Submit building permit including site plan with septic information
----- Prepare mini-contract to drill well and stake well location
----- Obtain well permit
----- Drill well
----- House stakeout
----- Phone for clearing
----- Flag for clearing
----- Clear lot
----- Order basement windows/doors
----- Phone for foundation excavation
----- Obtain building permit
----- Excavation layout
----- Phone for footings
----- Excavate foundation
----- Phone mason for block delivery
----- Deliver basement windows/doors
----- Footings
----- Phone for roof trusses
----- Phone for footing inspection
----- Phone for structural steel/drain tile/columns
----- Nail points in footings
----- Phone for underslab plumbing/laterals
----- Phone for concrete slab work
----- Deliver block
----- Build foundation walls
----- Phone for partial backfill
----- Phone carpenter
----- Phone for wall check survey
----- Order basement gravel
----- Phone for lst floor lumber package
----- Deliver structural steel
----- Parge foundation walls
----- Underslab plumbing
----- Wall check survey

----- Phone for parging inspection
----- Underslab plumbing inspection
----- Partial backfill
----- Grade basement subgrade and gravel
----- Pour basement floor
----- Inspection for slab
----- Grade garage subgrade
----- Dig and install W & S laterals or well and septic
----- Deliver lst floor lumber package
----- Set structural steel
----- Pour garage slab
----- Frame lst floor deck
----- Grade crawl space
----- Phone for 2nd floor lumber package/roof lumber
----- Phone for windows and exterior doors
----- Frame lst floor walls
----- Deliver 2nd floor lumber package/roof lumber
----- Frame 2nd floor deck
----- Phone for stair/rail measurement
----- Frame 2nd floor walls
----- Phone for brick veneer
----- Frame garage
----- Deliver roof trusses
----- Phone Pepco re: URD cable
----- Set roof trusses
----- Order shingles
----- Phone for plumbing rough-in
----- Phone for backfill and rough grade
----- Deliver windows and exterior doors
----- Phone heating and A/C rough-in
----- Set windows and exterior doors
----- Deliver stairs
----- Backfill and rough grade
----- Phone roofers
----- Phone for exterior trim package
----- Phone for exterior siding material
----- Begin brick veneer
----- Set interior stairs
----- Phone for siding installers
----- Roofing
----- Order hardware package
----- Plumbing rough-in
----- Phone for electric rough-in
----- Order hardwood flooring
----- Prepare interior finish schedule/door order
----- Order kitchen cabinets and vanities
----- Heating and A/C rough-in
----- Interior framing backup including soffits/box ducts

----- Phone for garage door
----- Phone for exterior trim
----- Deliver exterior trim package
----- Electric rough-in
----- Begin exterior trim (cornice)
----- Phone for concrete steps
----- Phone for insulation
----- Phone for drywall delivery and installation
----- Install water meter
----- Install furnace
----- Complete exterior trim (cornice)
----- Obtain electrical/plumbing rough-in inspections
----- Call for framing inspection
----- Framing inspection
----- Hang garage doors
----- Form concrete steps
----- ABC framing inspection
----- Frame punch-out
----- Pour concrete steps
----- Install insulation
----- Call for insulation inspection
----- Order light fixtures
----- Phone for interior door and trim package
----- Insulation inspection
----- Deliver drywall
----- Install electric meter
----- Phone for gutters and downspouts
----- Phone for drywall hanging
----- Turn on heat
----- Hang and tape drywall
----- Phone for driveway
----- Phone for exterior painting
----- Order ceramic tile
----- Hang gutters and downspouts
----- Exterior painting
----- Phone kitchen cabinet/vanities delivery
----- 1st coat drywall
----- Phone for final grading
----- Phone for rail delivery
----- Order kitchen appliances
----- Last date for paint selection
----- Deliver wood flooring
----- Lay wood floors
----- Deliver interior doors and interior trim
----- Interior trim
----- Final grade yard by excavator
----- Deliver kitchen cabinets/vanities
----- Install kitchen cabinets and vanities

----- Measure/order countertops
----- Phone for seeding or sodding
----- Install ceramic tile
----- Phone for switch and plug
----- Phone interior paint/caulk
----- Deliver hardware package
----- Prime paint
----- Drywall point-up/sand
----- Final grade yard for seed
----- Phone for final HVAC
----- Phone floor finishing
----- Deliver countertops
----- Set countertops
----- Deliver kitchen appliances
----- Sand and finish floors
----- Phone final survey
----- Paper hardwood floors
----- Deliver electrical fixtures
----- Hang electrical fixtures
----- Final HVAC/start-up/grills
----- Set plumbing fixtures
----- Set kitchen appliances
----- Call for water test
----- Lay carpet/vinyl
----- Call for final clean
----- Hang electrical fixtures
----- Final survey
----- Phone for A/C or heat check
----- Set finish hardware
----- Hang mirrors
----- Hang window screens/grills
----- Phone for final clean
----- Install shoe mold
----- Final clean
----- Walk-thru with client
----- Call for final inspection
----- Final inspection
----- Issue final occupancy permit

Checklists

Checklists are tools whose primary function is to help you avoid mistakes. Checklists should be designed to be as all-encompassing as possible because their true purpose is to force the user to run through a list of items and thereby assure that every item is dealt with or at least considered. Checklists can be of all types, sizes, forms and uses. They also provide management with the assurance that even employees who are less experienced than may be desired will consider the items on the checklist when they are doing their job. In this way, management is assured that compliance with the checklist will lead to satisfactory quality and conformance with accepted standards.

One of the best uses of a checklist is at the end of a phase of construction, particularly if things will be done which make it more difficult to repair original mistakes. An example of this is the installation of drywall. A good walk-through checklist used before drywalling helps eliminate the problems of going back and installing missing or forgotten components which are no longer exposed for easy access.

Checklists also serve a very useful function in the administrative end of the business and should be used for such things as controlling long lead time and special order items. As a management tool, each checklist should include a signature block so that responsibility can be assigned.

Remember, a good checklist that is used properly can save you from backtracking to make costly and time-consuming repairs, ensure that work is progressing in a planned fashion, keep employees on task and serve as a reminder to order materials for the next phase.

ABC BUILDERS
123 Main Street
Hometown, MD 21234
(301) 231-1234

Walk Lot: _____ _____

Prepare Specs: _____ _____

Review Specs: _____ _____

Prepare Costs: _____ _____

Review Cost: _____ _____

Prepare Contract _____ _____

Review Contract: _____ _____

Pre-commitment: _____ _____

Submit SDP: _____ _____

SDP Complete: _____ _____

Final Plan Review: _____ _____

Call Architect: _____ _____

Meet Architect: _____ _____

Pre-Dwg. Review: _____ _____

Final Drawing: _____ _____

Apply for Bldg.Permit: _____ _____

ABC BUILDERS
123 Main Street
Hometown, MD 21234
(301) 231-1234

Exterior
Roof sheathing   _____
Gable vents   _____
Rake and gutter boards   _____
Crickets and valleys   _____
Skylight curbs/cutouts   _____
Door sets   _____

Structural
Beam pockets   _____
Column plumb   _____
Joist hangers   _____

Garage
Garage door height & width   _____
Blocking ready for garage door   _____
Access panel or pull-down stairs per specs   _____

Kitchen
Bulkheads   _____
Square   _____
Blocking for wall cabinets   _____

Bathrooms	**Full**	**Half**	**Master**
Access panel	_____	_____	_____
Medicine chest	_____	_____	_____
5'-0" for tub	_____	_____	_____
Soffit bulkhead	_____	_____	_____
Special framing at tub, i.e. steps, soap shelf	_____	_____	_____
Blocking at tub corners	_____	_____	_____
Spread (12") at tubs	_____	_____	_____
Blocking for towel bars (48")	_____	_____	_____
Blocking for toilet paper	_____	_____	_____
Holder (16")	_____	_____	_____
Square at counter tops	_____	_____	_____

Stairs
Head height   _____
3'1" at bottom of basement stair   _____

ABC BUILDERS
123 Main Street
Hometown, MD 21234
(301) 231-1234

Technical

Joist size & spacing _____
Footing details _____
Beam sizes spans _____
Flitch plates _____
Duct chases _____

Windows & Doors

Exterior doors - r.o. - swing – brand _____
Windows - size - r.o. _____
Skylights – size – brand – operable _____
Fixed glass _____
Special – stained – transoms - side lites _____

General Framing

Mirrors _____
Medicine chests _____
Access panels (tubs) _____
Access panels (attic) _____
Access panels (garage) _____
Pull-down stairs _____
Laundry chutes _____
Towel bars _____
Toilet paper _____
Kitchen bulkhead _____

Electrical

Lighting layout _____
Surface mount _____
Recessed - all progress _____
Chain-hung _____
Fans _____
Track light _____
Oustide _____
Spots _____

Fixtures _____

Pole lamp/circuit _____

Driveway _____

Special _____

Outlets _____

Standard _____

Special _____

Kitchen _____

TV _____

Wall units _____

Basement (3 std) _____

Phone _____

Door chimes _____

Intercom _____

Speaker wire _____

Plumbing
Kitchen

Single bowl/double bowl _____

Dishwasher location _____

Disposal right or left sink bowl _____

Kitchen faucet w/ sprayer _____

Instant hot _____

Ice maker _____

Powder Room

Vanity _____

Bath

Vanities _____

Shower _____

Shower heads _____

Tubs - right or left hand _____

Jacuzzi _____

Other

Water heater location(s) size(s) _____

Pump tank location _____

Hose bibs (3 std) _____

Laundry room _____

Wall box _____

Laundry tray _____

Other special plumbing _____

Bars _____

Central vacuum locations _____

HVAC

Type and size system _____

Ducts by ASHRAE? _____

Returns _____

Central _____

Wall _____

Hi/low _____

Furnace location _____

Compressor location _____

Humidifier _____

Air cleaner _____

Condensate pump/drain _____

Thermostat _____

Day-night setback _____

Location _____

ABC BUILDERS
123 Main Street
Hometown, MD 21234
(301) 231-1234

Pre-construction Carpentry Meeting

BASEMENT
Beams
Pockets   _____

Splices   _____

Column lengths   _____

Block
Walls heights   _____

Step downs   _____

Double top plates   _____

Rough openings   _____

Finished portion of basement
Walls   _____

Stud height   _____

WINDOW & DOOR OPENINGS
Joists - 1st deck pkg
Size   _____

Direction   _____

Openings   _____

Stairs   _____

Hearth   _____

Duct chases   _____

Doubles for loads above   _____

Double for interior partitions (County Code?)   _____

Detail at step down at beam   _____

Step downs   _____

Cantilevers   _____

FIRST FLOOR
Plywood
Thickness   _____

Fir   _____

Built-up floors   _____

Exterior walls

Thickness   _____

Height   _____

Sheathing   _____

Rack bracing   _____

Special heights   _____

Header height   _____

Interior walls

Load-bearing: identify   _____

Flush headers   _____

Drop headers   _____

Interior door type and rough opening   _____

Bifold   _____

Bipass   _____

Knee walls

Joists

Size   _____

Direction   _____

Openings   _____

Stairs   _____

Hearth

Duct chases   _____

Cantilevers   _____

Step downs   _____

KITCHEN

Bulkheads: height/plan   _____

Wall cabinet blocking   _____

SECOND FLOOR

Plywood   _____

Thickness   _____

Exterior walls

Thickness   _____

Height   _____

Header height   _____

Special heights   _____

Interior walls

Load-bearing   _____

Flush headers   _____

Interior door type   _____

Knee walls   _____

Openings   _____

Medicine chests   _____

Tub access   _____

Attic access   _____

Showers   _____

Walls   _____

Seats   _____

ROOF

Truss layout

Overhang   _____

Cantilever   _____

Truncated   _____

Openings

Skylights   _____

Chimneys   _____

Ridge vent   _____

Gable vent   _____

Flat ceiling   _____

Identify locations   _____

Vaulted ceiling   _____

Truss   _____

Scissor   _____

One way   _____

Conventional   _____

Load-bearing ridge   _____

Collar ties   _____

Conventional   _____

Rafter sizes: spacing   _____

Ceiling joists:size/spacing   _____

Ridge beams: size   _____

Hip rafters: size   _____

STRUCTURAL
Microlams   _____

Doubles   _____

Triples   _____

Flitch plates   _____

Cornice detail   _____

Size   _____

Material   _____

Cedar   _____

Painted   _____

Wrapped   _____

Fypon   _____

Venting   _____

Build out for brick   _____

Attached to 2nd floor walls   _____

Rakes   _____

Gutter boards   _____

WINDOWS & DOORS
Window/Door schedule
Door type   _____

Window type   _____

Sizes   _____

Rough openings   _____

Header heights   _____

Color   _____

Circle heads   _____

Transoms   _____

Special   _____

Trapezoids   _____

Fixed panels   _____

Other   _____

REVIEW OF CROSS SECTIONS   _____

Pre-Drywall Checklist

One of the most critical stages of any custom house is the one just prior to applying the drywall. One day everything is exposed and visible and the next day it is covered up and inaccessible. One day it is easy to find and correct a mistake; the next day, it is impossible to find anything and expensive to fix the problem.

Most builders have some method of inspecting the house prior to drywall. The form presented in this document can be used to formalize the pre-drywall inspection process. By listing all of the items that need to be inspected, it ensures that nothing is overlooked. If you have several superintendents, the form can be used to promote a consistent approach to each inspection.

The Pre-Drywall Walk-Through Checklist covers the four critical areas of concern -- framing, electrical, HVAC and plumbing. We recommend that you use this checklist on a trial basis. Compare it to what you use now. Try it out on several jobs. Modify it or expand upon it to make it fit the specific needs of your business.

ABC BUILDERS**Pre-Drywall Checklist**
123 Main Street
Hometown, MD 21234
(301) 231-1234

FRAMING

General - The pre-drywall walk through should cover the following general framing items for all rooms:
- _____Backers for drywall
- _____Transfer of load from beams, headers, flitch plates and girder trusses through the wall system, including solid blocking to basement beams or masonry walls
- _____Size and spacing of floor joists as per plans
- _____Size and bearing for exterior wall headers
- _____Size and bearing for interior headers and multiple floor joists including hangers
- _____Bowed or bad studs for drywall especially in baths and kitchens
- _____Interior door openings

Garage -
- _____Garage door height and width
- _____Blocking for garage door tracks
- _____Access panel or pull-down stairs
- _____Beams and columns boxed for drywall
- Comments_____
  _____
  _____

Basement; Unfinished -
- _____Basement stairs set and framed
- _____All mechanicals stubbed through
- Comments _____
  _____
  _____

Kitchen -
- _____Bulkheads/light valances
- _____Blocking for wall cabinets
- _____Check square for "L" countertops
- Comments _____
  _____
  _____

Bathrooms -
- _____Tub framing tight to tub ends
- _____Medicine chest
- _____Access panel

- _____Bulkheads
- _____Tub platforms and steps
- Comments _____
  _____
  _____

ELECTRICAL

General - Check compliance with wiring diagram including location of lights, switches and plugs to code requirements.

Heavy Wiring -
- _____Dryer
- _____Heating/AC
- _____Attic HVAC
- Comments

- _____Pull-down iron
- _____Steamer/sauna
- _____Other _____

Comments _____
  _____
  _____

Kitchen -
- _____Wiring for all appliances
- _____Range hood
- _____Disposal
- _____Microwave

- _____Range
- _____Refrigerator
- _____Instant hot
- _____Trash compactor

- _____JennAir
- _____Dishwasher
- _____Wall oven

Low Voltage -
- _____Intercom
- _____Exterior speakers
- _____Telephone
- _____Television
- _____Speaker wires

- _____Master
- _____Door speakers

- _____Interior speakers

HVAC

General - Supply and return registers in proper location including alignment of ceiling registers, floor registers clear of doorways and circulation areas, wall returns clear of crown molding and wall cabinets.

Air Handles -
- _____Thermostat wires
- _____Condensate lines
- _____Drip pan
- _____Access panel and walkway to attic unit
- _____Refrigerator lines

Ventilation -
- _____Vents for all bath fans
- _____JennAir
- _____Range hood
- _____Wall oven
- _____Dryer
- _____Block for whole-house fan

General - Check tub(s) model and color and valves against specifications.
- _____Supply line material correct
- _____Drain line material correct
- _____Vent line material correct
- _____Supply and drain lines in outside walls or garage adequately insulated
- _____Ice maker lines
- _____Hose bibs
- _____Specialty sinks - bar, kitchen salad, work rooms, etc.

Sample Order Forms

Sometimes it seems the most commonly used order form is a scrap piece of 2 x 4. Let's face it -- we have all done it -- and know that it's wrong.

WHY USE ORDER FORMS?

- Make sure you don't forget a critical item

- Permanent written record of order

- Document when an order was placed, with whom and the promised delivery date

- Use in conjunction with purchase order system

- Save time by not rewriting the same type of order over and over

- Consistent record for other members of your organization if you are not available

- ORGANIZATION!

This section of the document presents Sample Order Forms for general items, masonry, structural steel, framing lumber, roof lumber, exterior windows and doors and interior trim. Use them as guides for more efficient and accurate ordering.

ABC BUILDERS
123 Main Street
Hometown, MD 21234
(301) 231-1234

General Order Form

JOB NAME _____

ORDERED BY _____

ORDERED FROM _____

DATE OF ORDER _____

NAME & TELEPHONE NUMBER _____

DELIVERY DATE _____

DESCRIPTION OF ORDER: _____

_____
_____
_____
_____
_____
_____
_____

ABC BUILDERS
123 Main Street
Hometown, MD 21234
(301) 231-1234

Masonry Block Order Form

JOB NAME: _____

DATE ORDERED: _____

ORDERED FROM: _____

12" Hollow Core	_____ Lightweight	_____ Heavyweight
12" FHA Caps	_____	
8" Hollow Core	_____	
8" FHA Caps	_____	
4" Solid 4's 4" x 8" x 16"	_____	
4" Hollow Core	_____	
Other Block	_____	
Other Block	_____	
Other Block	_____	
4" x 8" x _____ Lintels	_____	
4" x 8" x _____ Lintels	_____	
4" x 8" x _____ Lintels	_____	
4" x 8" x _____ Lintels	_____	Other _____
4" x 8" x _____ Lintels	_____	Other _____
Bags of Mortar Type S	_____	Brand Name _____
Bags of Mortar Type N	_____	Brand Name _____
Bags of Mortar Portland	_____	
14" anchor straps	_____	
Brick ties	_____	
Bundles 8" horizontal reinforcing	_____	
Bundles 12" horizontal reinforcing	_____	
Other _____	_____	
Other _____	_____	
Other _____	_____	

First Load	Second Load	Third Load	Fourth Load	Fifth Load
_____	_____	_____	_____	_____

ABC BUILDERS
123 Main Street
Hometown, MD 21234
(301) 231-1234

Structural Steel Order Form

JOB NAME: _____
DATE ORDERED: _____
ORDERED FROM: _____

BEAMS

Beam #	Size	Length	Special Information

FLITCH PLATES

Size	Hole Pattern	Bolts

LOLLY COLUMNS

Quantity	Size	Adjustable From	To

ANGLE IRONS

Quantity	Size	Length

MISCELLANEOUS STEEL/OTHER

Rebar

pcs. _____   size _____   length _____

_____   _____   _____

_____   _____   _____

6 x 6   Wire mesh _____ rolls
10 x 10   Wire mesh _____ rolls

<u>Other</u>   <u>Description</u>

_____

_____

_____

_____

_____

ABC BUILDERS
123 Main Street
Hometown, MD 21234
(301) 231-1234

Framing Lumber Order Form

JOB NAME: _____

DATE ORDERED: _____

ORDERED FROM: _____

(____) FIRST DECK   (____) SECOND DECK

_____ Rolls sill sealer

_____ 2 x 6 x 16' pressure treated

_____ 2 x 4 x 16' pressure treated

_____ 2 x _____ x _____ 1st Floor Walls - HEIGHT: _____

_____ 2 x _____ x _____ 1st Floor Walls - SIZE: 2x_____

_____ 2 x _____ x _____ 2nd Floor Walls - HEIGHT: _____

_____ 2 x _____ x _____ 2nd Floor Walls - SIZE: 2x_____

_____ 2 x _____ x _____ Garage Walls - HEIGHT: _____

_____ 2 x _____ x _____ Garage Walls - SIZE: 2x_____

_____ 2 x _____ x _____

_____ 2 x _____ x _____

_____ Sheets Underlayment T&G @ _____" thick _____ Pine _____ Fir

_____ 2 x 4 x 16'

_____ 2 x 4 x _____

_____ 2 x 4 x _____

_____ 2 x _____ x precut 104 5/8" (dbl. top plate) 9' wall

_____ 2 x _____ x precut 92 5/8" (dbl. top plate) 8' wall

_____ 2 x 4 x precut 94 1/4" (single top plate) 8' wall

_____ 2 x 6 x _____

_____ 2 x 6 x _____

_____ Sheets insulating sheathing at _____" thick
_____ R Max _____ Styrofoam _____ Other

_____ Sheets structural sheathing at _____" thick
_____ Celotex _____ CDX Plywood _____ Other

_____ Universal corner hangers

_____ Single joist hangers for 2 x _____

_____ Double joist hangers for 2 x _____

_____ Typar rolls

_____ Cases of construction adhesive

ABC BUILDERS
123 Main Street
Hometown, MD 21234
(301) 231-1234

Roof Lumber Order Form

JOB NAME: _____

DATE ORDERED: _____

ORDERED FROM: _____

_____ Sheets roof sheathing @_____" thick

    _____ Pine   _____ Fir   _____ Oriented strand board

_____ Boxes plyclips _____"

_____ Lin. ft. _____ x _____ facia   _____ material

_____ Lin. ft. _____ x _____ rake   _____ material

_____ Sheets _____" thick soffit   _____ material

_____ # Galvanized finish nails   _____ size

_____ 2 x _____ x _____ ridge rafters

_____ 2 x _____ x _____

_____ 2 x _____ x _____

_____ 2 x _____ x _____

_____ 2 x _____ x _____

_____ 2 x _____ x _____

_____ 2 x _____ x _____

_____ 2 x _____ x _____

_____ 2 x _____ x _____

_____ Single joist hangers

_____ Double joist hangers

ABC BUILDERS
123 Main Street
Hometown, MD 21234
(301) 231-1234

JOB NAME: _____
DATE ORDERED: _____
ORDERED FROM: _____

WINDOWS/SKYLIGHTS

Brand Name _____ Type _____ Color _____
Low E ( ) Yes ( ) No   Screens ( ) Yes ( ) No
Grilles ( ) Yes ( ) No

Description	Quantity	Location	Rough Opening	Comments

Description	Quantity	Location	Rough Opening	Comments

ABC BUILDERS
123 Main Street
Hometown, MD 21234
(301) 231-1234

JOB NAME: _____
DATE ORDERED: _____
ORDERED FROM: _____

TRIM

Quantity	Model #	DescriptionType	Location
_____	_____	BASE	FJ CLR
_____	_____	BASE	FJ CLR
_____	_____	BASE	FJ CLR
_____	_____	CASING	FJ CLR
_____	_____	CASING	FJ CLR
_____	_____	CASING	FJ CLR
_____	_____	CHAIR RAIL	FJ CLR
_____	_____	CHAIR RAIL	FJ CLR
_____	_____	CROWN MOLD	FJ CLR
_____	_____	PANEL MOLD	FJ CLR
_____	_____	PANELING	
_____	_____	OAK NOSING	
_____	_____	OAK CAP	
_____	_____	WINDOW STOOL	FJ CLR
_____	_____	INSIDE CORNER	
_____	_____	OGEE	
_____	_____	HANDRAIL	
_____	_____	HANDRAIL	
_____	_____	HANDRAIL BRACKETS	
_____	_____	STAIR RAILS & BALLISTERS CLOSET POLE	
_____	_____	MANTELS	
_____	_____	CLOSET POLE	
_____	_____	SHELF SUPPORTS	
_____	_____	PRS OF POLE ENDS	
_____	_____	NOVAPLY	
_____	_____	NOVAPLY	
_____	_____	1 X 3	
_____	_____	1 X 4	
_____	_____	1 X 6	
_____	_____	1 X 8	
_____	_____	1 X 10	
_____	_____	1 X 12	
_____	_____	BUNDLE SHIMS	
_____	_____	1/4 A/C PLYWOOD	

| | | UNDERLAYMENT _____ THICK | | |
| | | SCREWS | | |

Quantity	Model #	Description	Location	Type
_____	_____	PASSAGE LOCKS		
_____	_____	PRIVACY LOCKS		
_____	_____	EXTERIOR		
_____	_____	DEAD BOLTS		
_____	_____	DEAD BOLTS		
_____	_____	POCKET DOOR HDW		
_____	_____	MIRROR		
_____	_____	MIRROR		
_____	_____	MIRROR		
_____	_____	MIRROR		
_____	_____	MEDICINE CHEST		
_____	_____	MEDICINE CHEST		
_____	_____	MEDICINE CHEST		
_____	_____	TOWEL BAR		
_____	_____	TOWEL BAR		
_____	_____	TOILET PAPER		

INTERIOR DOORS

Size	Swing	Type	Casing	Location
_____	_____	_____		FJ CLR
_____	_____	_____		FJ CLR
_____	_____	_____		FJ CLR
_____	_____	_____		FJ CLR
_____	_____	_____		FJ CLR
_____	_____	_____		FJ CLR
_____	_____	_____		FJ CLR
_____	_____	_____		FJ CLR
_____	_____	_____		FJ CLR
_____	_____	_____		FJ CLR
_____	_____	_____		FJ CLR
_____	_____	_____		FJ CLR
_____	_____	_____		FJ CLR
_____	_____	_____		FJ CLR
_____	_____	_____		FJ CLR
_____	_____	_____		FJ CLR

EXTERIOR DECKS

Quantity	Pressure Treated
_____	
_____	2 X 4 X _____
_____	2 X 4 X _____
_____	2 X 2 X _____
_____	4 X 4 X _____
_____	6 X 6 X _____
_____	2 X 6 X _____
_____	2 X 8 X _____
_____	2 X 10 X_____
_____	2 X 10 X_____
_____	2 X 12 X_____
_____	2 X 12 X_____

12 D GALVANIZED COMMON _____ #
10 P GALVANIZED FINISH _____ #

LAG BOLTS/WASHERS _____

_____

_____

CONTRACT PACKAGE

Contract and Associated Documents

No document is more important to the relationship between the builder and the client than the contract and the related contract documents (specifications, allowance schedule, and draw schedule). Above all else, it is imperative that the contract be fair to both parties and obviously interpreted as such by both parties. There is no better sales agent for you, the custom builder, than the attorney of the client, when he states to the client, "This is an eminently fair contract and it is surprising to see such from a builder."

Every contract should contain the following:

1. The name of the parties involved, the location of the work and a very brief description of the work, e.g., construction of a custom residence located at _____.

2. The time of commencement of the work and the approximate time of substantial completion. A phrase which is used in the example states "that due to normal variations of the custom home process, substantial completion may be 45 days earlier or later than projected." This type of up-front acknowledgement that a firm date may not be totally possible will go a long way toward alleviating disputes and misunderstandings later on.

3. The contract sum should be clearly stated, both numerically and in written form.

4. The concept of progress payments must be delineated. Specific references should be made to the deposit as well as the allowable time after due notice by contractor within which progress payments are due and payable.

5. A provision for interest on late payments must be included in every contract.

6. A reference to change orders, and the time which payment for such change orders is due must be included.

7. Provisions for final payment must be made within the contract.

8. Reference to associated contract documents -- the plans, the specifications, the allowance schedule, the draw schedule and the warranty -- must be specifically included.

9. The obligations of the owner should be clearly delineated. These obligations include timely selection by owner of all materials, colors and allowance items, as well as a blanket statement that the owner agrees to cooperate and make every reasonable effort to assist the contractor.

10. The obligations of the contractor should also be clearly delineated.

11. Subcontracts should be defined within the contract and any criteria governing such subcontracts should be clearly listed.

12. A provision for the settling of disputes within the contract must be included. An arbitration clause is one alternative which may be used. Such clauses are enforceable in court, and almost always will be upheld by the appropriate jurisdiction. While many people believe that arbitrators often tend to "cut the baby in half," in many instances an arbitration is a much simpler and less expensive procedure than a court case. Additionally, many arbitrators tend to use less stringent standards than may be employed by the courts, frequently to the builder's favor.

13. References to delays caused by the owner through change orders or by unavoidable circumstances should be referenced as a possible justification for delay in completion.

14. References to contractors' liability insurance as well as builders' risk insurance should be made within the contract, specifically stating the obligations of all parties and clearly denoting responsibilities for obtaining insurance coverage.

15. References to changes in the work, i.e., change orders, must be made within the contract. It is imperative that a statement that all such changes shall be authorized in writing be in the contract.

16. Provision for termination of the contract should be included. It is this author's opinion that the owner should be given the unilateral power to terminate the contract assuming that he reimburses the contractor for all labor and materials furnished plus a proportional share of the overhead and profit associated with the job. While many believe that this gives too much power to the owner, I believe strongly that this clause will rarely be used, because the owner really has no other place to go, and additionally, if circumstances deteriorate to the point where the owner wishes the contractor off the job, it is probably best for all parties concerned if he leaves.

17. In fairness to the client, it is appropriate to include a contingency for permanent financing in an amount appropriate for the construction of the residence.

BUSINESS BUILDER

A contract is a living thing. It should constantly be updated and improved. And when you decide something should be added (or deleted), don't hesitate. Do it.

123 Main St.
Hometown, MD 21231
(301) 231-1234

FIXED SUM CONTRACT

This is a binding legal contract between the parties. If you do not fully understand the contract you should consult an attorney.

1. PARTIES
   A. Date – This contract between the parties is made as of (_____date____)
   B. Owner – The contract is between (_____Owner_____), hereafter referred to as the Owner and;
   C. Contractor – (_____Contractor____), hereafter referred to as the Contractor.

2. SCOPE OF WORK
   A. Work – This contract is for the (__description of work__)
   B. Location – The location of the work to be performed under this contract is (_____location of work_____)
   C. Contract Documents – The work to be performed under this contract shall be governed by the documents enumerated in Paragraph 5 of this contract which shall become part of this contract.

3. TIME
   A. Start Date – The work to be performed under this contract shall start on (_____start date____).
   B. Duration – The work shall be substantially complete within _____ months after the start date. This date shall be the completion date.
   C. Delay of Start – In the event the start date is delayed for more than 45 days due solely to the Owner or to conditions pertaining to the Owner's lot, then the contract amount shall be increased by _____% on the 46th day and by _ _____% for each subsequent 30 days thereafter.
   D. Cancellation – If the start date is delayed more than 180 days due solely to the Owner or to conditions pertaining to the Owner's lot, then the contract is null and void and any deposit monies shall be refunded to the Owner.
   Penalty Clause – If the work is not substantially complete within 30 days after the completion date, then the contractor shall pay the Owner $_____ per day until substantial completion is obtained.
   Time is of the Essence – Time is of the essence in this contract and the Contractor shall manage the work to ensure achievement of substantial completion by the completion date.
   Substantial Completion – The date of substantial completion is when construction is sufficiently complete so that the Owner can fully utilize the work for its intended use. The date of substantial completion shall not precede the date of final

occupancy permit or the date of the final inspection approval by the lending institution.

Time Changes – Changes in any times in this contract shall be made only by change order signed by both parties to the contract

1. If the Contractor is delayed by labor disputes, fire, unusual delays in transportation, unusual weather conditions or other conditions beyond the control of the Contractor, then the parties may mutually agree upon an extension of times by change order.
2. If changes are requested by the Owner, then the times may be extended by mutual agreement of the parties.

4. CONTRACT AMOUNT
   A. Fixed Sum – The Contractor shall complete the work to be performed under this contract for the fixed sum of (\$____contract amount____).
   B. Deposit – The Owner shall pay the Contractor (\$_____deposit amount) upon the signing of this contract. This deposit shall be subject to any provisions of this contract pertaining to deposits.
   C. Draws – The Owner shall pay the Contractor in accordance with the approved draw schedule.
      1. Requests – The Contractor shall request draw payments when work required by the draw schedule is completed.
      2. Payments – Draw payments shall be made to the Contractor within 3 days after approval by the lending institution in the full amount approved by the lending institution. Note: If a lending institution is not used, the functions of the lending institution shall be assumed by a mutually agreed upon third party.
      3. Interest – Any overdue payments under this contract shall bear interest at a rate of (____annual interest rate %) per year.
      4. Change Orders Due – Any payments due for change orders shall be due with the next draw payment after completion of the change order work.
      5. Allowance Overage – If the Owner selects materials whose cost is in excess of 115% of the specified allowance, the Contractor shall prepare an allowance overage change order.

Change Order – The Owner may request changes to the work.
1. During Construction – If the change is requested during construction, the Contractor shall prepare a cost and time estimate, including overhead and profit at a rate of _____%. The change shall be implemented upon approval of the change order by the Owner.
2. Allowance Accounting – Upon completion of an allowance category, the Contractor shall provide an accounting to the Owner, including overhead and profit at a rate of _____% if allowance costs exceed 115% of the allowance amount.

Final Payment
1. Due – Final payment is due within 3 days after substantial completion.
2. Temporary Occupancy – When weather, seasonal or other job related conditions prevent a final occupancy permit from being issued, then a temporary occupancy permit shall meet the requirements of substantial completion. The Contractor shall pay for any bonds, fees, etc. required to obtain a temporary occupancy permit.

3. Occupancy – The Owner shall not occupy the work until temporary or final permits are obtained and until final payment has been made to the Contractor.
4. Walk-Through – When a request for final payment is made, the Owner and Contractor shall conduct a walk-through inspection. All mutually agreed upon items shall be listed. The dollar value to be withheld by the Owner for each item shall be mutually agreed upon by the Owner and Contractor. These amounts shall be paid to the Contractors as soon as each individual item has been completed.
5. Waiver of Liens – At the date of substantial completion, the Contractor shall provide a waiver of liens statement indicating that all subcontractors and suppliers have been paid or will be paid from the proceeds of the final draw. The Contractor shall be liable for any lien claims, including any cost and reasonable attorney's fees required to discharge any liens.
6. Arbitration – Any disputes or claims arising out of the request for final payment shall be settled by arbitration in accordance with Paragraph 11 B of this contract.

Termination

1. If an arbitrator employed in accordance with Paragraph 11 B determines that the Contractor defaults or persistently fails or neglects to carry out the Work in accordance with the Contract Documents or fails to perform any provision of the Contract, the Owner, after seven days' written notice to the Contractor and without prejudice to any other remedy he may have, may make good such deficiencies and may deduct the cost thereof from the payment then or thereafter due the Contractor or may terminate the Contract and take possession of the site and of all materials, equipment, tools, and construction equipment and machinery thereon owned by the Contractor and may finish the Work by whatever method he may deem expedient, and if the unpaid balance of the Contract Sum exceeds the expense of finishing the Work, such excess shall be paid to the Contractor, but if such expense exceeds such unpaid balance, the Contractor shall pay the difference to the Owner.

2. The Owner may terminate this Contract for reasons other than provided in Paragraph 4.F.1 by providing ten (10) days' written notice to the Contractor. In such event the Contractor shall be reimbursed from construction financing funds and funds of the Owner for all work completed prior to notice of termination. The total amount of money to be paid to the Contractor shall be determined using the construction draw schedule as a basis for computing the money due for the work completed prior to notice of termination. Any and all deposit monies shall be retained by the Contractor.

3. In the event of a loss greater than 30% of contract value due to occurrences covered by property insurance described in Paragraph 8.C, the Contractor shall be paid from combined proceeds of insurance, construction financing, and Owners funds, for all work completed prior to notice of termination. The total amount of money to be paid to the Contractor shall be determined using the construction draw schedule as a basis for computing the money due for the work completed prior to notice of termination. Any and all deposit monies shall be retained by the Contractor. At this time the original contract shall be automatically terminated. At Owners option Contractor agrees to negotiate in good faith a new contract for the repair and/or completion of the custom residence.

5. In the event of a loss less than 30% of contract value due to occurrences covered by property insurance described in Paragraph 8.C, the Contractor shall prepare a written change order for the labor, materials, and profit and overhead costs required to repair the damage caused by the occurrence. Upon acceptance of the change order by the Owner, Contractor shall proceed to repair the damages caused by the occurrence. Cost of the change order shall be paid from proceeds of insurance, and funds of the Owner. Subsequent to completion of the repairs, Contractor shall continue work under the original contract. Should Owner choose not to accept the change order as prepared by Contractor, the provisions of Paragraph 4.F.3 shall apply.

6. CONTRACT DOCUMENTS
   A. The contract documents listed below are incorporated as part of this contract and in total are the entire agreement between the parties.
   Plans – signed on   _____
   Specifications – signed on   _____
   Allowance Schedule – signed on   _____
   Draw schedule – signed on   _____
   Warranty – signed on   _____
   B. In the event of a conflict between the plans and specifications, the specifications shall govern work not covered in these contract documents will not be required unless it is reasonable to assume that the work is necessary to the completion of the project.

7. RESPONSIBILITIES OF CONTRACTOR
   A. Site Inspection
      1. Contractor acknowledges that he has inspected the site and is aware of the existing conditions.
      2. Barring any other reference to testing herein or contained in the specifications, subsurface conditions not obvious to visual inspection are not the responsibility of the Contractor.
   B. Supervision
      1. Contractor is solely responsible for determining the methods, and order of construction, and for supervising and coordinating all segments and phases of the work.
      2. Contractor shall utilize and employ persons and subcontractors who are competent in the tasks assigned. Discipline and accountability shall be maintained by the Contractor.
      3. The Contractor shall abide by all rules, ordinances, or regulations governing the execution of the work. Any deviations or variances of the plans or specifications for such rules and ordinances shall be brought to the attention of the Owner immediately.
      4. Contractors shall routinely maintain the cleanliness of the site of the work. When the work is complete, all waste materials as well as all equipment, tools or excess materials shall be removed.

B. Materials
   1. Unless otherwise specified or agreed to, all of the materials and equipment included in the work shall be new. All such materials will be as specified, free from defect and of good quality.
   2. Contractor shall pay for all materials including sales, use, or other similar taxes which are in effect at the time this contract is executed. Additionally, unless otherwise specified, the Contractor shall pay for any and all government permits or fees required to accomplish the work.

C. Labor
   1. The Contractor shall provide and pay for the labor and services necessary to accomplish the work in a timely fashion.
   2. The Contractor shall be responsible for any acts and omissions of his employees and subcontractors.
   3. To the fullest extent of the law, the Contractor shall hold the Owner harmless from all claims arising out of the performance of the work, when such claims are:
      a. Attributable to bodily injury or death, or injury to or destruction of tangible property (not including the Work itself) and;
      b. Caused in whole or in part by a negligent act or omission of the Contractor, his employees, or his Subcontractors. This obligation shall not relieve, negate or abridge any other right of indemnification which would otherwise exist. This indemnification obligation is not limited in any way by any restriction, or limitation on the type or amount of benefits or compensation payable under any relevant worker's compensation act.

D. Subsurface Conditions
This contract assumes standard excavation with no subsurface solid rock, streams or springs which require excavation or special equipment, other than a tracked front loader. Excavation costs above standard conditions are to be paid as an extra. This contract assumes subsurface soil bearing conditions adequate for the footing sizes shown on plans. Additional footing costs are to be paid as an extra.

E. Protection of Persons and;
The Contractor shall be responsible for initiating, maintaining and supervising all safety precautions and programs in connection with the work. He shall take all reasonable precautions for the safety of, and shall provide all reasonable protection to prevent damage, injury or loss to (1) all employees on the work and other persons who may be affected thereby, (2) all the work and all materials and equipment to be incorporated therein, and (3) other property at the site or adjacent thereto. The Contractor shall remedy all damage or loss to any property caused in whole or in part by the Contractor his employees or any Subcontractor, except damage or loss attributable to the acts or omissions of the Owner or his agents or by anyone for whose acts the Owner may be liable, and not attributable to the fault or negligence of the Contractor.

F. Repair and Correction of Work

The Contractor shall promptly correct or repair any work defective or failing to conform to the Contract Documents whether observed before or after substantial completion: The Contractor shall correct any work found to be defective or nonconforming within a period of one year from the date of substantial completion of the contract or within such longer period of time as may be prescribed by law or by the terms of any applicable special warranty required by the contract documents.

8. RESPONSIBILITIES OF OWNER
   A. The Owner shall furnish all surveys and record plats and a legal description of the building site.
   B. Except as provided in Paragraph 6C, the Owner shall secure and pay for necessary approvals, easements, assessments and charges required for the construction, use or occupancy of permanent structures.
   C. The Owner shall forward all instructions to the Contractor.
   D. The Owner shall sign all necessary applications required to obtain permits and bonds.
   E. The Owner shall select, in a timely manner, all allowance items, materials and colors required during the construction process.
   F. The Owner shall obtain all necessary approvals, and/or acknowledgements from any Architectural Board or Committee whose jurisdiction is relevant to this project.
   G. Owner agrees to cooperate and make every reasonable effort to assist Contractor and further agrees to permit Contractor to place signs on or about the premises during the course of construction and allow Contractor to show the home to potential customers during the course of same. Owner shall not contract with, authorize or permit, for any reason whatsoever, any individual, other contractor or subcontractor to perform, in whole or in part, any additions or changes to the plans and specifications unless authorized to do so by the Contractor in writing. Owner shall not communicate directly with any workman, employees, agents or subcontractors of the Contractor, unless so directed by the Contractor.

9. INSURANCE
   A. Contractor shall obtain and maintain liability insurance in the amount of $_____ to protect him from workers' compensation claims and claims for damages because of bodily injury, including death, and from claims for damages, other than to the work itself, to property which may arise out of or result from the Contractor's operations under this contract, whether such operations be by himself or by any employee or Subcontractor. This insurance shall be written for not less than any limits of liability specified in the Contract Documents, or required by law, whichever is the greater.
   B. The Contractor shall be responsible for purchasing and maintaining his own liability insurance and, at his option, may maintain such insurance as will protect him against claims which may arise from operations under the contract.
   C. The Contractor shall purchase and maintain property insurance upon the entire work at the site to the full insurable value thereof. This insurance shall include

the interests of the Owner, the Contractor and Subcontractors in the work and shall insure against the perils of fire and extended coverage and shall include "Builder's Risk" insurance for physical loss or damage including, theft, vandalism and malicious mischief. The insurance shall include glass coverage with a maximum deductible of $_____ per occurrence.

D. Any loss insured under Paragraph 7C is to be adjusted with the Owner and made payable to the Owner as trustee for the insured.

E. The Owner and Contractor waive all rights against each other for damages caused by fire or other perils to the extent covered by insurance obtained pursuant to this contract or any other property insurance applicable to the work, except such rights as they may have to the proceeds of such insurance held by the Owner as trustee.

10. SUBCONTRACTS

A. Definition – A Subcontractor is a person or entity who has a direct contract with the Contractor to perform any of the work at the site.

B. List of Principal Subcontractors – The Contractor, as soon as practicable, shall furnish to the Owner in writing the names of Subcontractors for each of the principal portions of the work.

C. Reasonable Objections to Subcontractors – The Contractor shall not employ any Subcontractor to whom the Owner may have a reasonable objection. The Contractor shall not be required to contract with anyone to whom he has a reasonable objection.

D. Required of Subcontractors – Contracts between the Contractor and the Subcontractors shall; (1) require each Subcontractor, to the extent of the work to be performed by the Subcontractor, to be bound to the Contractor by the terms of the Contract Documents, and to assume toward the Contractor all the obligations and responsibilities which the Contractor, by these documents, assumes toward the Owner and, (2) allow to the Subcontractor the benefit of all rights, remedies and redress afforded to the Contractor by these Contract Documents.

11. OTHER

A. Governance – This contract shall be governed by the Laws of the State of

_____.

B. Arbitration

Unless the parties mutually agree otherwise, all claims or disputes between the Contractor and the Owner arising out of, or relating to, the Contract Documents or the breach thereof shall be decided by arbitration in accordance with the current Construction Industry Arbitration Rules of the American Arbitration Association. Notice of the demand for arbitration shall be filed in writing with the other party to the Owner-Contractor Agreement and with the American Arbitration Association and shall be made within a reasonable time after the dispute has arisen. The award rendered by the arbitrators shall be final, and judgement may be entered upon it in accordance with applicable law in any court having jurisdiction thereof.

C. Contingencies – This contract is null and void in the event Owner is unable to obtain construction financing in the amount of $_____ at an interest rate not to exceed _____%.

12. SIGNATURES

This contract shall constitute the sole and entire agreement between the parties whose signatures are affixed below.

_____   _____

Owner   Date

_____   _____

Owner   Date

_____   _____

ABC BUILDERS   Date

123 Main St.
Hometown, MD 21231
(301) 231-1234

COST PLUS FIXED FEE CONTRACT

This is a binding legal contract between the parties. If you do not fully understand the contract you should consult an attorney.

1. PARTIES
    A. Date – This contract between the parties is made as of (_____date_____)
    B. Owner – The contract is between (_____Owner_____), hereafter referred to as the Owner and;
    C. Contractor – (_____Contractor_____), hereafter referred to as the Contractor.

2. SCOPE OF WORK
    A. Work – This contract is for the (___description of work___)
    B. Location – The location of the work to be performed under this contract is (_____location of work_____)
    C. Contract Documents – The work to be performed under this contract shall be governed by the documents enumerated in Paragraph 5 of this contract which shall become part of this contract.

3. TIME
    A. Start Date – The work to be performed under this contract shall start on (_____start date_____).
    B. Duration – The work shall be substantially complete within _____ months after the start date. This date shall be the completion date.
    C. Delay of Start – In the event the start date is delayed for more than 45 days due solely to the Owner or to conditions pertaining to the Owner's lot, then the contract amount shall be increased by _____% on the 46th day and by _____% for each subsequent 30 days thereafter.
    D. Cancellation – If the start date is delayed more than 180 days due solely to the Owner or to conditions pertaining to the Owner's lot, then the contract is null and void and any deposit monies shall be refunded to the Owner.
    E. Penalty Clause – If the work is not substantially complete within 30 days after the completion date, then the contractor shall pay the Owner $_____ per day until substantial completion is obtained.
    F. Time is of the Essence – Time is of the essence in this contract and the Contractor shall manage the work to ensure achievement of substantial completion by the completion date.
    G. Substantial Completion – The date of substantial completion is when construction is sufficiently complete so that the Owner can fully utilize the work for its intended use. The date of substantial completion shall not precede the date of final

occupancy permit or the date of the final inspection approval by the lending institution.

   H. Time Changes – Changes in any times in this contract shall be made only by change order signed by both parties to the contract

     1. If the Contractor is delayed by labor disputes, fire, unusual delays in transportation, unusual weather conditions or other conditions beyond the control of the Contractor, then the parties may mutually agree upon an extension of times by change order.

     2. If changes are requested by the Owner, then the times may be extended by mutual agreement of the parties.

4. CONTRACT AMOUNT

   A. Contractor's Fee – The Contractor shall be paid a fixed fee of ($ contract amount   ) to supervise the work to be completed under this contract.

   B. Deposit – The Owner shall pay the Contractor ($ deposit amount   ) upon the signing of this contract. This deposit shall be subject to any provisions of this contract pertaining to deposits.

   C. Draws of Contractor's Fee – The Owner shall pay the Contractor (# of payments) equal consecutive monthly payments of ($ monthly payment   ), starting on (1st payment date) with the final payment payable upon project completion.

Changes – The Owner may request changes to the work..

     1. There shall be no increase in the Contractors fee for changes to the plans and specifications.

     2. For any major additions to the plans such as outbuildings, tennis court, swimming pool, etc., an increased fee shall be negotiated between the Owner and Contractor.

Costs to be Reimbursed or Paid Directly by Owner

     1. The term cost of the work shall include costs set forth below incurred in the proper performance of the work and paid by the Contractor or directly by Owner.

     2. Wages paid for labor in the direct employ of the Contractor in the performance of the work including welfare, unemployment compensations, social security and other benefits.

     3. Cost of all materials, supplies and equipment incorporated in the work, including costs of transportation thereof. All discounts for cash or prompt payment shall accrue to the Owner.

     4. Payments made by the Contractor or Owner to Subcontractors for work performed pursuant to Subcontractors under this Agreement.

     5. Cost of all materials, supplies, equipment, temporary facilities and hand tools not owned by the workers, which are consumed in the performance of the work.

     6. Reasonable rental costs of all necessary machinery and equipment, exclusive of hand tools, used at the site of the work, whether rented from the Contractor or others.

     7. Cost of premiums for all job site bonds and job site insurance, permit fees and sales use or similar taxes related to the work.

8. Losses and expenses, not compensated by insurance or otherwise, sustained by the Contractor in commission with the work, provided they have resulted from causes other than the fault or neglect of the Contractor.
9. Cost of removal of all debris.

Costs Not to be Reimbursed

1. The term Cost of the Work shall not include any of the items set forth below.
2. Salaries or other compensation of the Contractor's personnel at the Contractor's offices.
3. Expenses of the Contractor's offices.
4. Any part of the Contractor's capital expenses, including interest on the Contractor's capital.
5. Costs due to the negligence of the Contractor, any Subcontractor, anyone directly or indirectly employed by any of them, of for whose acts any of them may be liable, including, but not limited to, the correction of defective or nonconforming work, disposal of materials and equipment wrongly supplied, or making good any damage to property.
6. Overhead, general expense, and the cost of any item not specifically or reasonably inferable as included in the items described in Paragraph 4.E.

Accounting Records – The Contractor shall check all materials, equipment and labor entering into the work and shall keep such full and detailed accounts as may be necessary for proper financial management under this Agreement. The Owner shall be afforded access to all the Contractor's records relating to this Contract.

Payments to the Contractor –

1. Based on Application for Payment submitted by the Contractor, the Owner shall make progress payments to the Contractor or directly to Subcontractors and suppliers as follows: On the _____ and _____ of each month the Contractor shall submit to the Owner invoices for direct payment to Contractor, suppliers and Subcontractors with accompanying statements. Said invoices shall be paid by the Owner on the _____ and the _____ of each month. Prior to submitting invoices to the Owner, the Contractor shall review the work progress to insure that it is in accordance with the plans and specifications. If deficiencies are found, the Contractor shall notify the suppliers and Subcontractors and recommend to the Owner an amount to be withheld from payment.
2. When approximately 50% of the Work has been completed the Contractor shall obtain partial Waiver of Liens from Subcontractors and suppliers paid directly by the Contractor. At the time of final payment, the Contractor shall obtain final Waiver of Liens in such form as approved by the Owner.

I. Final Payment – Final Payment of the Contractor's Fee is due within _____ days after the date of substantial completion as defined by issuance of a final or temporary Use and Occupancy Permit and when construction is sufficiently complete so that the Owner can occupy or utilize the Work for the use for which it is intended.

5. CONTRACT DOCUMENTS
  A. The contract documents listed below are incorporated as part of this contract and in total are the entire agreement between the parties.
  Plans – signed on   _____
  Specifications – signed on   _____
  Allowance Schedule – signed on   _____
  Warranty – signed on   _____
  B. In the event of a conflict between the plans and specifications, the specifications shall govern work not covered in these contract documents will not be required unless it is reasonable to assume that the work is necessary to the completion of the project.

6. RESPONSIBILITIES OF CONTRACTOR
A. Site Inspection
  1. Contractor acknowledges that he has inspected the site and is aware of the existing conditions.
  2. Barring any other reference to testing herein or contained in the specifications, subsurface conditions not obvious to visual inspection are <u>not</u> the responsibility of the Contractor.
B. Supervision
  1. Contractor is solely responsible for determining the methods, and order of construction, and for supervising and coordinating all segments and phases of the work.
  2. Contractor shall utilize and employ persons and subcontractors who are competent in the tasks assigned. Discipline and accountability shall be maintained by the Contractor.
  3. The Contractor shall abide by all rules, ordinances, or regulations governing the execution of the work. Any deviations or variances of the plans or specifications for such rules and ordinances shall be brought to the attention of the Owner immediately.
  4. Contractors shall routinely maintain the cleanliness of the site of the work. When the work is complete, all waste materials as well as all equipment, tools or excess materials shall be removed.
C. Materials
  1. Unless otherwise specified or agreed to, all of the materials and equipment included in the work shall be new. All such materials will be as specified, free from defect and of good quality.
  2. Contractor shall pay for all materials including sales, use, or other similar taxes which are in effect at the time this contract is executed. Additionally, unless otherwise specified, the Contractor shall pay for any and all government permits or fees required to accomplish the work.
D. Labor
  1. The Contractor shall provide and pay for the labor and services necessary to accomplish the work in a timely fashion.
  2. The Contractor shall be responsible for any acts and omissions of his employees and subcontractors.

3. To the fullest extent of the law, the Contractor shall hold the Owner harmless from all claims arising out of the performance of the work, when such claims are:
   a. Attributable to bodily injury or death, or injury to or destruction of tangible property (not including the Work itself) and;
   b. Caused in whole or in part by a negligent act or omission of the Contractor, his employees, or his Subcontractors. This obligation shall not relieve, negate or abridge any other right of indemnification which would otherwise exist. This indemnification obligation is not limited in any way by any restriction, or limitation on the type or amount of benefits or compensation payable under any relevant worker's compensation act.

E. Subsurface Conditions

This contract assumes standard excavation with no subsurface solid rock, streams or springs which require excavation or special equipment, other than a tracked front loader. Excavation costs above standard conditions are to be paid as an extra. This contract assumes subsurface soil bearing conditions adequate for the footing sizes shown on plans. Additional footing costs are to be paid as an extra.

F. Protection of Persons and;

The Contractor shall be responsible for initiating, maintaining and supervising all safety precautions and programs in connection with the work. He shall take all reasonable precautions for the safety of, and shall provide all reasonable protection to prevent damage, injury or loss to (1) all employees on the work and other persons who may be affected thereby, (2) all the work and all materials and equipment to be incorporated therein, and (3) other property at the site or adjacent thereto. The Contractor shall remedy all damage or loss to any property caused in whole or in part by the Contractor his employees or any Subcontractor, except damage or loss attributable to the acts or omissions of the Owner or his agents or by anyone for whose acts the Owner may be liable, and not attributable to the fault or negligence of the Contractor.

G. Repair and Correction of Work

The Contractor shall promptly correct or repair any work defective or failing to conform to the Contract Documents whether observed before or after substantial completion: The Contractor shall correct any work found to be defective or nonconforming within a period of one year from the date of substantial completion of the contract or within such longer period of time as may be prescribed by law or by the terms of any applicable special warranty required by the contract documents.

7. RESPONSIBILITIES OF OWNER
   A. The Owner shall furnish all surveys and record plats and a legal description of the building site.
   B. Except as provided in Paragraph 6C, the Owner shall secure and pay for necessary approvals, easements, assessments and charges required for the construction, use or occupancy of permanent structures.
   C. The Owner shall forward all instructions to the Contractor.
   D. The Owner shall sign all necessary applications required to obtain permits and bonds.

E. The Owner shall select, in a timely manner, all allowance items, materials and colors required during the construction process.

F. The Owner shall obtain all necessary approvals, and/or acknowledgements from any Architectural Board or Committee whose jurisdiction is relevant to this project.

G. Owner agrees to cooperate and make every reasonable effort to assist Contractor and further agrees to permit Contractor to place signs on or about the premises during the course of construction and allow Contractor to show the home to potential customers during the course of same. Owner shall not contract with, authorize or permit, for any reason whatsoever, any individual, other contractor or subcontractor to perform, in whole or in part, any additions or changes to the plans and specifications unless authorized to do so by the Contractor in writing. Owner shall not communicate directly with any workman, employees, agents or subcontractors of the Contractor, unless so directed by the Contractor.

8. INSURANCE

A. Contractor shall obtain and maintain liability insurance in the amount of $_____ to protect him from workers' compensation claims and claims for damages because of bodily injury, including death, and from claims for damages, other than to the work itself, to property which may arise out of or result from the Contractor's operations under this contract, whether such operations be by himself or by any employee or Subcontractor. This insurance shall be written for not less than any limits of liability specified in the Contract Documents, or required by law, whichever is the greater.

B. The Contractor shall be responsible for purchasing and maintaining his own liability insurance and, at his option, may maintain such insurance as will protect him against claims which may arise from operations under the contract.

C. The Contractor shall purchase and maintain property insurance upon the entire work at the site to the full insurable value thereof. This insurance shall include the interests of the Owner, the Contractor and Subcontractors in the work and shall insure against the perils of fire and extended coverage and shall include "Builder's Risk" insurance for physical loss or damage including, theft, vandalism and malicious mischief. The insurance shall include glass coverage with a maximum deductible of $_____ per occurrence.

D. Any loss insured under Paragraph 7C is to be adjusted with the Owner and made payable to the Owner as trustee for the insured.

E. The Owner and Contractor waive all rights against each other for damages caused by fire or other perils to the extent covered by insurance obtained pursuant to this contract or any other property insurance applicable to the work, except such rights as they may have to the proceeds of such insurance held by the Owner as trustee.

9. SUBCONTRACTS
    A. Definition – A Subcontractor is a person or entity who has a direct contract with the Contractor to perform any of the work at the site.
    B. List of Principal Subcontractors – The Contractor, as soon as practicable, shall furnish to the Owner in writing the names of Subcontractors for each of the principal portions of the work.
    C. Reasonable Objections to Subcontractors – The Contractor shall not employ any Subcontractor to whom the Owner may have a reasonable objection. The Contractor shall not be required to contract with anyone to whom he has a reasonable objection.
    D. Required of Subcontractors – Contracts between the Contractor and the Subcontractors shall; (1) require each Subcontractor, to the extent of the work to be performed by the Subcontractor, to be bound to the Contractor by the terms of the Contract Documents, and to assume toward the Contractor all the obligations and responsibilities which the Contractor, by these documents, assumes toward the Owner and, (2) allow to the Subcontractor the benefit of all rights, remedies and redress afforded to the Contractor by these Contract Documents.

10. OTHER
    A. Governance – This contract shall be governed by the Laws of the State of
    _____.
    B. Arbitration
    Unless the parties mutually agree otherwise, all claims or disputes between the Contractor and the Owner arising out of, or relating to, the Contract Documents or the breach thereof shall be decided by arbitration in accordance with the current Construction Industry Arbitration Rules of the American Arbitration Association. Notice of the demand for arbitration shall be filed in writing with the other party to the Owner-Contractor Agreement and with the American Arbitration Association and shall be made within a reasonable time after the dispute has arisen. The award rendered by the arbitrators shall be final, and judgement may be entered upon it in accordance with applicable law in any court having jurisdiction thereof.
    C. Contingencies – This contract is null and void in the event Owner is unable to obtain construction financing in the amount of $_____ at an interest rate not to exceed _____%.

11. SIGNATURES
    This contract shall constitute the sole and entire agreement between the parties whose signatures are affixed below.

    _____   _____
    Owner   Date

    _____   _____
    Owner   Date

    _____   _____
    ABC BUILDERS   Date

137

Allowance Schedule

All "allowance items" within the contract should be clearly defined on a separate schedule. The schedule should include a description and definition of the different types of allowances, specifically material-only allowances and installed allowances. The schedule should clearly state that material costs include not only the material itself, but taxes, shipping and delivery to the job site.

Each allowance item should be listed, and a brief description of that item as well as the amount of the allowance given. For example, carpet, which is often given as an installed allowance, should specifically state that it includes the carpet, pad, tack strip, transition strips and labor to install.

The most common allowance items among custom builders are as follows: cabinets, countertops and vanities, appliances, hardware, floor coverings (carpet, vinyl, ceramic tile), light fixtures, landscape, alarm systems and well and septic systems. In addition, items which are not totally allowances will often include a single allowance item. As an example, the cost of brick siding on a house may be included within the price of the contract, contingent upon the inclusion of a brick at a certain material allowance, such as $300 per 1,000 bricks. In this case, only the material used is an allowance item. Labor and installation have already been figured into the job.

The proper preparation and use of an allowance schedule can substantially reduce misunderstandings and conflicts between the client and the builder. Also, every allowance item removes the degree of uncertainty to the builder and protects against cost overruns in that particular category. The only danger of allowances is that if the builder is careless in his estimates or deliberately uses low allowances, this will result in large allowance overruns which tend to make most clients unhappy.

BUSINESS BUILDER
When properly used, allowances offer two significant advantages to the builder.

1. They permit the preconstruction process (bidding and contract document preparation) to proceed rapidly without undue delay caused by decisions in selecting materials.

2. They remove all cost overrun potential from the cost category in which they are used. Since many of these cost categories are done at the end of the job, they effectively protect cost categories with high potential exposure due to inflation and the natural tendency of materials to cost more tomorrow than today.

ABC BUILDERS
123 Main Street
Hometown, MD 21234
(301) 231-1234

Allowance Schedule

NOTES:

- All material costs include taxes, shipping and delivery charges to job site.

- Material Only Allowance: Cost of material delivered to site. Unless otherwise noted ABC is responsible for installation of the materials within the contract price.

- Installed Allowance: Cost of material delivered to site plus all labor charges required to install materials in house.

Description   Type of Allowance   Amount

CABINETS, COUNTERTOPS, VANITIES, MIRRORS,
SHOWER DOORS, MEDICINE CABINETS
Kitchen, bathrooms, bars, kitchen, bathrooms, bars.   Installed $_____

(If ABC installs cabinets and countertops, there will be a labor charge against the allowance of 10% of the total allowance expenditures. Installation of shower/tub doors will be billed at $45.00/door.)

HARDWARE AND ACCESSORIES
Towel bars, soap dishes, toilet paper holders, special
closet accessories, all interior and exterior door knobs
& locks.   Material $_____

APPLIANCES
Kitchen range, refrigerator, oven, dishwasher, trash
compactor, disposal, icemaker, others.   Material $_____

NOTE: Installation charges to include $75 for a wall oven and $50 for a cooktop. Appliance trim kits are also charged to this allowance. See cost on price sheet attached.

CARPET
Carpet, pad, tack strip, transition strips, labor.
                                               Installed $_____

VINYL
Underlayment, transition strips, labor and setting
materials.   Installed $_____

CERAMIC TILE
Labor, setting material, grout, accessories, marble
thresholds, underlayment, wainscoat.   Installed $_____

NOTE: Allow $25.00 per 4x8 sheet of 1/2" plywood underlayment for ABC to supply and
install. ABC, within the contract price, will provide and install water-resistant drywall and
wonder board.

LIGHT FIXTURES
Includes door chimes, interior and exterior fixtures,
ceiling fans, bathroom heat light fans.   Material $_____

(ABC will rough wire and install light fixtures for "standard" layout within the contract
price. Wiring and installation of fixtures above "standard" will be billed as a Change
Order for extra work. - See price sheet attached.)

LANDSCAPE AND GRASS
Fine grading, seeding, sodding, landscaping, stone
edgings, moving trees, tree wells to protect existing
trees, offsite topsoil, mulching, strawing, tree pruning,
dead tree removal, retaining walls.   Installed $_____

DECKS
Post footing, posts, docking, rail, benches/planters
and labor to install.   Installed $_____

INTERCOM
Master unit, remote units, wire, labor to install.   Installed $_____

CENTRAL VACUUM   Installed $_____

ALARM SYSTEM   Installed $_____

FRONT DOOR
Doors, jambs, weatherstripping, thresholds,
hanging charge if doors are not prehung.   Installed $_____

SEWER AND WATER CONNECTION FEES.	Fee	$_____
WELL (see Plumbing Specifications)	Installed	$_____
SEPTIC SYSTEM (See Plumbing Specifications)	Installed	$_____
ELECTRIC/TELEPHONE CONNECTION	Fee	$_____

DRIVEWAY APRON/CULVERT PIPES
CR-1 sub-base, CR-6 base, asphalt, concrete
or asphalt apron to County standards & culvert
pipe(s) if required   Installed $_____

BOOKSHELVES/BUILT-INS
All plywood, lumber, cabinet doors, countertops,
hardware and labor necessary to construct all
bookcases, bookshelves and built-ins. Cost of labor
and material to stain paint and seal same.   Installed $_____

AGREEMENT:

_____   _____
Owner   Date

_____   _____
Owner   Date

_____   _____
ABC Builders   Date

Specifications

It is imperative that detailed specifications be prepared for each residence constructed. While these specifications will be created from a master specification typically stored in a word processor, it is necessary for each and every client that these specifications be modified and customized to reflect as exactly as possible the specific materials included in the residence.

Specifications in many ways offer the contractor substantial protection from overruns and unexpected conditions, if provisions for these items are included within the specifications. Several examples are:

1. A rock clause requiring that such nonstandard excavation as rock or underground springs be billed as an extra.

2. An importing or exporting of fill clause stating that any dirt brought into or removed from the job shall be billed as an extra. (If some earth moving is anticipated, this may be established as an allowance, when the extras apply only to amount above the allowance.)

3. The use of allowances for those items which are indeterminate, or undefined at the time of the contract. Wells, kitchen cabinets, light fixtures and landscaping are just several examples of cost categories which lend themselves to the allowance approach.

ABC BUILDERS   Specifications
123 Main Street
Hometown, MD 21234
(301) 231-1234

FOR _____

1.   GENERAL
•   Plans and architectural services required to obtain a building permit are to be provided by the Contractor. Reproduction of additional sets is to be paid by the Contractor.

•   The Contractor shall obtain and pay for all permits and bonds required by the County for construction of the residence. These specifications assume that the Owner has a recorded building lot with approved septic or sewer connection and approved well or water connection. The Contract includes an allowance of $ _____ to drill, steel case, grout and test the well. All engineering and survey work required to obtain or modify an approved building lot shall be paid for by the Owner.

•   Builders' risk insurance in an amount not less than the Contract Sum shall be maintained for the duration of the Contract. In general, the policy shall provide all risk, damage and liability coverage to protect both the Owner and the Contractor. The cost of the insurance shall be paid by the Contractor.

•   Temporary utilities required in the construction of the house through substantial completion shall be provided for and paid by the Contractor. These shall include water, electricity and heat as required. The Owner shall be responsible for utilities after the time of substantial completion.

2. SITE PREPARATION
•   The Contractor shall stake the house location for final approval by the Owner. After the foundation is built, the Contractor shall provide a wall check survey. At time of substantial completion the Contractor shall provide a final survey. The cost of additional survey work required to locate and/or permanently mark property corners/lines shall be the responsibility of the Owner.

•   The Owner and Contractor shall mutually agree upon which trees are to be cut and what area is to be cleared. Hardwood trees will be cut into 2' lengths and stored on-site up to a max. of approximately three cords. Brush shall be chipped or hauled away. Tree stumps shall be hauled away. The Contractor is not responsible for the survival of uncut trees.

3. EXCAVATION
•   The Contractor shall excavate, provide standard backfill and rough grade to obtain positive drainage away from the house. Final fine grading in preparation for seeding/sodding shall be part of the landscape allowance.

- The following items are not included in the Contract and if required shall be billed as extras (1) nonstandard excavation such as rock or underground springs, (2) compaction of backfill, (3) additional excavation or soil treatment caused by inadequate soil bearing capacity, i.e., less than 2500 lbs. per square foot, (4) importing or exporting of fill dirt or top soil.

4. FOUNDATIONS

- Footings shall be 3000 psi 6 bag mix concrete. The size and reinforcement, if any, shall be in accordance with the plans.

- Foundation walls to be Concrete Masonry Units (CMU) with Dur-O-Wall at 16" on center (O.C.) vertically in 12" walls. The size and height of foundation walls as per plans. Basement block work is 14 courses. Extra block to be billed at $2.85/12"; $2.50/8"; $2.40/4" CMU.

- Waterproofing to be 1/2" parging coated with foundation tar, roller or spray applied.

- Structural steel as per plans, columns to be 3" outside diameter (O.D.).

- Pressure treated lumber shall be used for sill plates with sill sealer and anchor straps 6'- 0" O.C. minimum.

- Foundation drain shall be 4" perforated plastic pipe covered with clean gravel and 15# felt paper. Drain to grade outside basement foundation. See Plumbing Specification for more detail.

- Radon protection measures shall consist of a 4" perforated plastic pipe on the inside of the basement wall vented to a vertical stack.

- Basement floor to be 4" concrete with 6x6 #10 wire mesh over 4" gravel and 6 mil polyethelene vapor barrier.

- Termite protection shall include a termite inspection and chloradane poisoning with a 5-year warranty, if permitted by Howard County.

- Garage floor to be 4" concrete with 6x6 #10 wire mesh and 6 mil polyethelene vapor barrier. At the garage door the concrete will be turned down to a depth of 18" or there will be a continuous footing provided. For a standard size garage the floor will be sloped 3" for drainage. If fill dirt is required in the garage the slab will be reinforced with #4 rebars.

5. UTILITIES

- Septic field and tank shall be installed in accordance with county requirements. System to be sized with/without garbage disposal. Septic allowance amount per allowance schedule. See Plumbing Specifications for more detail.

- The Owner shall dig, case, grout and test a well in the approved location. The Contractor shall provide a connection from the well to the house. See Plumbing Specifications for more details.

- The Contractor shall have electric and telephone service installed from the main lines to the house. The Contractor shall have the electric meter installed. The allowance amount for BG & E or PEPCO to provide the service connection is $_____.

- Main electric service to be 300 amp.

- The house shall be prewired for three (3) telephone outlets. Connections to telephone outlets shall be the responsibility of the Owner. Addt'l. outlets at $30 ea.

- House shall be prewired for Two (2) TV antenna outlets, wires to terminate in attic for future antennas. Co-axial cable shall be used. At the Owners request, the wires can terminate for connection to cable TV service.

6. FRAMING
- Floor joist size, spacing and grade as per plans. If no grade is shown #2KD Hem-Fir is the standard.

- Exterior walls 2x4's 16" o.c. with double top plate. Studs to be KD-SPF. Headers over all windows and doors in load bearing walls to be double 2"x12" unless shown otherwise.

- EXTERIOR WALL HEIGHT TO BE: _____

- FIRST FLOOR WALL HEIGHT TO BE: _____

- SECOND FLOOR WALL HEIGHT TO BE: _____

- Roof framing to be manufacturers approved trusses and/or conventional framing with size, spacing and grade as per plans.

- Subflooring to be 3/4" T&G fir plywood glued and nailed to the floor joists. 1/2" underlayment under ceramic floors. 1/4" masonite under vinyl floors.

- Wall sheathing to be 1/2" R-Max with plywood corners for rack bracing.

- Roof sheathing to be 1/2" CDX fir plywood with 1/2" plyclips.

- Exterior deck material to be pressure treated No. 2 Southern Yellow Pine (SYP) with No. 1 2x6 used for decking.

7. ROOFING
- Certainteed Glassguard shingles applied over 15# felt, with aluminum valley and step flashing.

- Gutters and downspouts to be seamless aluminum in standard colors of either brown, beige or white. Splashblocks to be provided at terminal of downspouts.

8. EXTERIOR SIDING
- Exterior brick, as per plans with a $240/thousand allowance for standard size brick materials including tax and delivery. Brick to be laid with standard gray mortar, struck and washed.

- Exterior of house to be wrapped in Tyvek, or equivalent, prior to application of siding.

- Exterior siding to be _____.

- Exterior soffit detail to be _____.

9. EXTERIOR WINDOWS AND DOORS
- Exterior windows to be Anderson double-hung WHITE with insulated glass and screens. All oval and round windows and skylites to be by Andersen Windows. Size and location as per plans. Window grilles not included.

- Exteriors doors per Schedule D.

10.   INTERIOR WALLS
- HANGING - House will be hung with 1/2" regular drywall on ceiling and walls. Shower walls and bath tub walls will be hung with 1/2"-water resistant board. The bottom three feet of all showers shall be hung with Wonder Board or equivalent. Garage ceiling and walls will be hung with 1/2" firecode* drywall. Contract includes drywall down the basement stairs on one side of the stud walls. The remainder of the basement is unfinished unless otherwise specified on the plans and specifications.

- Meets local codes.

- GLUE AND SCREWING - House will be screwed in uniform manner with applicable screws. Drywall adhesive will be applied to every stud and ceiling joist before application of drywall.

- FINISHING - Finishing of drywall takes three steps. You receive: 1) one tape coat. This is tape that covers all seams, joints and corners; 2) block coat, this is the second coat used to cover the tape; and 3) the skim coat used to taper out any imperfections of drywall mud in two previous coats. The entire house will then be sanded smooth with sandpaper. After primer coat the house will then be pointed up, i.e., walking through house with a bright light, shining it on walls and ceilings looking for imperfections and fixing them. When house is finished, Owner will walk through and approve final product or point out any place which in their opinion still needs attention.

- PAINTING AND STAINING
  PAINT: Walls and trim will receive two coats of Duron Pro-Kote paint. First coat consisting of spray applied primer and second coat which is roller and brush applied. All

trim, with exception of closet and linen shelves, will be semi-gloss and walls will be flat. Kitchen and baths can be done in semi-gloss entirely if client chooses. When custom colors are used, if they do not cover primer with one coat, a charge of $ 0.25 per sq. ft. will be made for third coat in room.

STAIN: Wood will be spot sanded where necessary. Stain will be rag or brush applied. Sander sealer coat will then be applied. Light sanding and/or steel wool will be done before final coat is put on. Final coat will be polyurethane.

11. INTERIOR FINISH
- Kitchen cabinet and counter tops per allowance schedule.

- Vanities, mirrors, and shower/tub doors per allowance schedule.

- Stairs from first to second floor to be oak treads, oak risers, oak ballisters, oak handrail, and painted pine stringers. Stairs to basement to be No.1 pine treads and risers and pine stringers. Stair size, and location and railing detail to be as per plans.

- Interior doors in accordance with Schedule A.

- Closet shelving to be 1x12 or 1x16 Novaply - painted.

- Interior trim in accordance with Schedule B.

- Baseboard - Every room in house will have baseboard. Carpet will be tucked to this where applicable. Where there is either wood floor, ceramic floor, or vinyl floor a shoe mould will be used.

- Trim closet - master bedroom - completed according to Owner's plans. Other closets to have one (l) pole and one (l) shelf at proper height with closet support used where span is greater than 4'. Linen closets to have five (5) shelves, four (4) 16-inch shelves and one (1) 12-inch shelf on top.

12. ELECTRICAL
- Standard wiring included in the Contract is as follows:
- Receptacles installed per code. - Two (2) double floods - Two (2) exterior receptacles - Twelve (12) recessed lights - One (1) water heater connection - One (1) switch receptacle in each bedroom - One (1) doorbell circuit - One (1) paddle fan - Well pump as required - Bsmt: Two (2) receptacles and four (4) keyless - Garage: One (1) receptacle, two (2) lights, garage door opener plugs - Appliances: refrigerator, dishwasher, disposal, cooktop/oven, wall oven with microwave - HVAC as per plans - Extra boxes to be installed at a cost of $30 each. - Lighting fixtures to be charged against the lighting allowance include addt'l. recessed fixtures, bathroom fans, fans, exterior lights, landscape lights, pole lights. Charge to hang extra light fixtures is $30 each.

NOTES: Three-way circuits will be provided at stairways, rooms with two entrances, front door and one (1) additional door selected by Owner.

UNLESS OTHERWISE NOTED IN THE SPECIFICATIONS WIRING FOR THE FOLLOWING ITEMS IS NOT INCLUDED IN THE CONTRACT:

Instant hot, separate microwave, freezer, trash compactor, exterior lights not attached to house, i.e., pole lamp or driveway lights, pool wiring, attic fan, whole house fan, pull-down iron, whirlpool, steamers, four-way switches, dimmers, floor receptacles, under-cabinet lights, assembly of tracklights and/or chandeliers, sump pumps, condensate pumps, heat light fans, closet lights other than walk-in closets.

13. HVAC
• Heating and air conditioning heat pump to be two (2) systems, Carrier or equal. Size of system and number and location of supplies and returns to be determined by HVAC Contractor.

• Optional extras are Carrier 49WS humidifier @ $265 each, Honeywell F50A electric air cleaner @ $550 each at time of rough-in or $650.00 at final, Honeywell night setback thermostat @ $225 each.

14. PLUMBING
• BASEMENT ROUGH-IN
Future drain lines and vent stack connection will be provided at the finished concrete floor level for the bathroom configuration shown on the plans. If the grade permits, the drain lines will be connected to the septic tank (or sewer). If the basement grade is above the septic tank (or sewer) elevation, then a sewage ejector pump tank will be installed. If required, the additional cost of the ejector pump tank will be $ 250.00. The ejector pump, its installation at a future date and electrical connection are not included in the contract. Water supply lines will be provided in the basement at a point that is accessible to the future bathroom. The contract does not include any stud walls, electrical outlets or bathroom fans and/or fan connections.

• WATER CONNECTION - WELL
The contract includes the installation of a Gould 1/2 hp two wire, 220 volt or equivalent, submersible well pump suitable for wells to a depth of 240 feet. For deeper wells up to a depth of 340 feet, a Gould 3/4 hp two wire, 220 volt or equivalent submersible well pump will be used at an additional cost of $200.00. The pump will be connected to the house with a 1" Selflex or poly pipe for wells within 100' of the house. Additional distance from the house will be billed at a rate of $2.50 per foot.

• WATER CONNECTION - PUBLIC WATER
The contract includes the installation of a 1" poly pipe from the public water house connection valve up to a distance of 100 feet. Additional distance from the house will be billed at a rate of $2.50 per foot. The contract does not include any water "tap" fees or the installation of the public water house connection valve unless an allowance is specifically included in the allowance schedule.

- ## SEWER CONNECTION - SEPTIC

The installation of the septic system is an allowance item per the allowance schedule. The system will be sized in accordance with County requirements for a _____ bedroom house with/without a garbage disposal. The type of system required may vary including drywell, leeching bed, deep trenches or shallow trenches, and will be determined in conjunction with the County. The size and depth of the field is based on the soil conditions and groundwater level. The labor and material items whose installed cost constitute the septic field allowance include: septic tank, distribution box, pipe, stone, filter paper, block for drywells, lids, covers and cleanouts, trenching, excavation, and backfill.

- ## SEWER CONNECTION - PUBLIC SEWER

The contract includes the installation and connection of a 4" PVC pipe from the house to an existing sanitary sewer house connection whose location is marked by a visible wood marker. The maximum distance from the house to the sewer connection within the contract price is 50 feet. Additional cost for longer distances will be $9.00 per foot. The maximum depth of the sewer at any point between the house and the connection is assumed to be 12 feet. Additional depth will be billed as a change order. The contract does not include any sewer "tap" fees.

- ## WATER HEATER

The contract includes A.O. Smith - 5 yr. warranty Energy Miser, or equivalent water heater(s). If two heaters are provided, the water distribution will be approximately balanced between supply points in the house. If oil or gas heaters are used they will be vented as per manufacturers requirements.

- ## WELL PUMP PRESSURE TANK

A V140 Gould Aqua-Air, or equivalent model pressure tank will be installed.

- ## SUPPLY LINES

Water supply lines will be C PVC plastic or schedule M copper. The size of the lines, unless otherwise specified, will be determined by the plumbing subcontractor. There will be one main supply cut-off valve accessible in the basement and individual cut-off valves to other fixtures. There will be _____ outside frostproof hose bibs located as per the plans. If not specified on the plans, they will be distributed one each to the front, rear and garage parking areas.

- ## WASTE AND VENT LINES

Waste and Vent Lines will be PVC and their location and size will be determined by the plumbing subcontractor. Whenever possible, vent stacks will be located on the rear of the roof. Options available to the Owner are: 1) main vertical drain located in 6-inch wall with insulation to reduce noise.

- ## BASEMENT DRAIN LINES

If the house has a walk-out basement a 2" PVC drain line will be run to daylight. This line will receive discharges of HVAC condensate, water heater emergency overflow and humidifier emergency overflow. If there is no walk-out basement then a sump pit with pump will be provided to receive above mentioned discharges. Options available to the Owner are: HVAC condensation pump for direct discharge to outside $145.00 each.

- ## BASEMENT DRAINS

If the house has a walk-out basement a 4 inch diameter perforated plastic drain pipe, discharging to daylight will be used. The pipe will be covered with 1/2 inch diameter gravel and 15 lbs. roofing paper in accordance with County requirements and will be installed on the outside of the Foundation. If there is no walk-out, the 4 inch diameter pipe will be installed on the interior of the foundation and will drain to a plastic pump pit fitted with a submersible sump-pump. All basement areaway drains will drain to the interior sump pit. Options available to the Owner for clay soil conditions and/or low lying areas are: outside drain to daylight and inside drain to sump pit with pump $260.00.

- ## PLUMBING FIXTURES

Location/Type	Standard
Kitchen	
Sink	Double Bowl Stainless Steel Republic HT
Sink Faucet	Delta 300 w/sprayer
Instant Hot	None
Disposal*	See Appliance Allowance/Selection
Icemaker*	Standard connection
Island sink	None
Island sink faucet	None
Other	
Other	
Powder Room	
Toilet	Kohler Wellworth EB
Toilet fixtures	Chrome fittings
Sink	Kohler Pennington Oval
Sink faucets	Delta 2522 chrome - 2 handle
Other	
Basement Bath	
Tub	Kohler Villager
Tub faucets	Delta 636 chrome w/spout & shwr head
Shower	Ceramic* over vinyl pan
Shower faucets	Delta 622 chrome
Sink	Kohler Pennington Oval
Sink faucets	Delta 2522 chrome
Toilet	Kohler Wellworth EB
Toilet fixtures	chrome
Ist Floor Bath	
Tub	Kohler Villager
Tub faucets	Delta 636 chrome

Shower	Ceramic* over vinyl pan
Shower faucets	Delta 622 chrome
Sink	Kohler Pennington Oval
Sink faucets	Delta 2522 chrome
Toilet	Kohler Wellworth EB
Toilet fixtures	chrome

2nd Floor Bath

Tub	Kohler Villager
Tub faucets	Delta 636 chrome w/spout & shwr head
Shower	Ceramic* over vinyl pan
Shower faucets	Delta 622 chrome
Sink	Kohler Pennington Oval
Sink faucets	Delta 2522 chrome
Toilet	Kohler Wellworth EB
Toilet fixtures	chrome

Master Bath

Tub	Kohler Villager
Tub faucets	Delta 636 chrome w/spout & shwr head
Shower	Ceramic* over vinyl pan
Shower faucets	Delta 622 chrome
Sink	Kohler Pennington Oval
Sink faucets	Delta 2522 chrome
Toilet	Kohler Wellworth EB
Toilet fixtures	chrome
Bidet	
Steamer	

NOTES
* Icemaker, disposal, and ceramic tile are allowance items.
* All Kohler tubs, sinks and toilets are assumed to be in standard colors.

15. INSULATION

- Wall insulation to be R-13.

- Vaulted ceiling insulation to be R-19 or R-30 depending on depth of rafter with baffles if required.

- Crawl space insulation to be R-19.

- Exposed above-grade basement walls at walk-out to be insulated with R-8 in accordance with County requirements.

- Ceiling insulation to be R-30.

- Basement band joist to be insulated with R-13.

16. FINISH FLOORS

- Carpet per allowance schedule.

- Ceramic tile per allowance schedule.

- Vinyl per allowance schedule.

- Finish floors based on Schedule C.

- Hardwood: 2' 3/4" strip red oak, select grade, installed, sanded, stained one coat, sealed and one coat polyurethane.

17. EXTERIOR GROUNDS/PATIOS/WALKS

Grade-
- Final finish grade, seeding, sodding, fountains, and plantings shall be installed per allowance schedule.

Driveway-
12' wide driveway with 20' x 35' parking apron consisting of 4" to 6" of crusher run stone and 2" asphalt paving. Location as per revised site plan. Allowance per allowance schedule.

Patios-
- All patios per plans. If no patios are shown, each exterior door at grade shall have a broomed concrete pad, 4" thick, sized as follows:
  Width - Door + 1' on each side   Depth - (Distance from house) - 4'

- Elevated doors shall exit to wood decks per the allowance schedule.

Walks-
- All walks shall be broomed concrete, per plan. If no walks are shown, contract includes 90 sq. feet of concrete walkway, 4" thick.

18. APPLIANCES

- Appliances per allowance schedule.

19. HARDWARE

- Kitchen and vanity hardware installed as part of allowance.

- Interior locks to be Schlage F-Series _____ style, _____ finish _____.

- Exterior locks to be Schlage F-Series.

- Exterior deadbolts to be Schlage single or double cylinder.

- Front door lock allowance for standard bore lock $ _____.

- Each bathroom will receive one (1) Miami Carey Classica Chrome #7006 toilet paper holder and two (2) #7010-24" towel bars.

- Polished brass door stops, hinges, and pulls on interior doors.

20. FIREPLACE

* Fireplace is located in plan. Finishes as follows:

SCHEDULE A - INTERIOR DOORS

lst/2nd Floors................... 6-panel hollow core   masonite
Basement......................... NOT IN CONTRACT

SCHEDULE B - INTERIOR TRIM

Room	Base	Windows	Doors	Chair Rail	Crown	Other
Living						
Dining						
Others						

NOTE: All numbers refer to (Millwork Co. of ABC Builders choice) specifications.

SCHEDULE C - FINISH FLOORS

Location	Type of Floor
LIVING RM	
DINING RM	
FOYER	
KITCHEN	
LAUNDRY	
FAMILY RM	
STAIRS	
POWDER RM	
MASTER BEDRM	
BEDRM #1	
BDRM #2	
BDRM #3	
BDRM #4	
UPSTAIRS BATH	
OTHER:	
OTHER:	
OTHER:	

SCHEDULE D - EXTERIOR DOOR SCHEDULE

LOCATION   MATERIAL   MODEL

ACCEPTANCE:

_____   _____
Owner   Date

_____   _____
Owner   Date

_____   _____
ABC Builders   Date

Charges/Prices

To improve organization and minimize the redundancy of constantly repricing the same changes requested by clients, it is a good idea to have a Standard Prices sheet. This should list the most common changes requested by your clients and be periodically updated. In addition, it should be shown to and discussed with your clients before construction begins, thus setting the stage for two important concepts:

1. There will be changes, clients will add items to their construction costs, and you are going to charge them for these changes.

2. The cost of common items will be established so the clients have some understanding of what things cost and do not live under the illusion that small additional items tend to be cheap or free.

You may wish to add a signature block following the Standard Price Sheet so the clients can sign, thus acknowledging the cost of changes. This will avoid any future discussions or arguments over charges for the items on the list. If you wish, the list may also include more general prices such as hourly rates for different skill levels and personnel. Again, in this way, should a change require the time of an electrician or finish carpenter, for example, there will be no debate when the clients are billed at the previously disclosed hourly rate.

ABC BUILDERS
123 Main Street
Hometown, MD 21234
(301) 231-1234

Add 2 ft. increments to house length	(per sq. ft.)	$ 46.00
	(per sq. ft./brick)	$ 49.00
Add 2 ft. increments to garage length	(per sq. ft.)	$ 20.00
Add 2 ft. increments to garage length	(per sq ft./brick)	$ 23.00
Decks (plus stairs, benches, and planters)	(per sq. ft.)	$ 11.00
Screened in porch	(per sq. ft.)	$ 40.00
Humidifier		$ 300.00
Electronic air cleaner		$ 600.00
Attic fan		$ 260.00
Whole house fan		$ 295.00
Intercom (includes Master unit, front door & 3 add'l. speakers)		$1,025.00
Recessed lights (complete, w/housing, trim, bulbs)		$ 90.00
Basebord heaters (installed) (see electrical price list)		
Garage door openers (each)		$ 300.00
Hardwood flooring (select-oak-stained)	(per sq. ft.)	$ 5.40
Bleaching (pickling is additional charge)	(per sq. ft.)	$ 1.50
Provide/install ISE H330 instant hot faucet		$ 190.00
Upgrade 52 gallon water heater to 66 gallon		$_____
Upgrade 66 gallon water heater to 80 gallon		$ 125.00
Basement rough-in (gravity feed)		$ 475.00
Additional hose bibs (each)		$ 100.00
Skylights (Andersen) - 24x48 each		$ 650.00
Pull-down stairs (each)		$ 175.00
Basic alarm system		$1,985.00
Deluxe central vac system (five (5) outlets, all accessories)		$1,525.00
Install standard shower or tub doors	(each)	$ 45.00
Install trim kit to refrigerator door	(labor only)	$ 50.00
Install trim kit to dishwasher	(labor only)	$ 25.00
Reverse refrigerator doors - if required		$ 30.00
Retaining walls - (wood) Per 6"x6"x8' p.T. Pc		$ 35.00
Per 6"x6"x12' p.T. Pc		$ 45.00
Dirt Incoming (from ABC/to ABC) per load		$ 50.00
Outgoing (from ABC/to ABC) per load		$ 20.00
Molding: 1 pc. Crown - #49	(per linear ft.)	$ 3.25
2 Pc. Crown - #49/663	(per linear ft.)	$ 5.00
1 Pc. Chair - #390	(per linear ft.)	$ 2.75
Shelving 12"	(per linear ft.)	$ 4.00
Shelving 16"	(per linear ft.)	$ 4.60
Metal chimney caps (all sizes) installed		$ 85.00

I understand that the prices shown above apply for any items listed which I request in my new home and which are not specifically included in my Contract and Specifications. Additionally, for any item not prepriced, ABC BUILDERS agrees to provide a price. Upon approval by the client, this price will constitute a change order as stated in the Contract.

ACCEPTANCE

Signature _____ Date _____
Owner _____

Design Contract

It's paperwork time! Alas, the days of a handshake and verbal contracts are long gone.
Every custom builder must have a complete and workable set of contract documents.
This section of the package presents two forms of contract documents. The first is a
sample Design Contract and the second is a written Policy for Owners Using Their Own
Subcontractors or Suppliers.

The Design Contract is an essential part of every custom builder's package.

The advantages of using a design contract include:

- Defines each step of the design process

- Gets client to commit to each successive stage of design

- Specifies design costs up front

- Ensures that the custom builder gets compensated for creating the design even if
the client builds with someone else.

- Acts as a marketing tool to keep the client working with you as a professional

ABC BUILDERS   **Design Contract**
123 Main Street
Hometown, MD 21234
(301) 231-1234

Agreement Between Owner and Contractor for Design Contract Where Basis
of Payment is a Stipulated Sum

AGREEMENT made as of this _____ day of _____ in the year
of Nineteen Hundred and _____ between the Owner, _____ and
ABC Builders, hereafter referred to as the Contractor, for Design of a Custom Residence
located at _____.

ARTICLE I - Purpose

1.1   ABC Builders is a full service custom builder that specializes in the design and
construction of custom homes. The purpose of this Design Contract is to set forth and
clarify the four (4) phases of the design process and to agree to a fixed price for the
completion of each of these phases.

1.2   In the event that the Contractor enters into a Contract to build the house for the
Customer, then all costs paid for in the Design Contract shall be credited to the
Customer.

ARTICLE II - Phase 1 <u>Product Definition</u>

2.1   During Phase 1, the Contractor and Customer will define the house that will be
designed during Phases 2 - 4.

2.2   The following activities will occur during Phase 1:

•   Initial meeting with the Customer to collect all existing information about the
proposed house such as floor plans, brochures and ideas that the Customers already
have.

•   Review of house requirements including size, number and type of rooms, style,
number of stories, basement and garage requirements. The Contractor will utilize
existing plans to guide the Customer in defining house requirements.

•   Establish a target budget including land costs, house costs, financing and soft
costs.

•   Review the Customers' building lot, including a field visit as required, to determine
the initial feasibility of constructing the house on the lot.

- If no lot is owned by the Customer, the Contractor will assist the Customer in screening potential lots and in providing information on lots for sale in the location and price range desired by the Customer.

- The Contractor will provide the Customer with information on construction and permanent financing and assist the Customer in making contact with suitable lending institutions.

- Establish a time table for completion of Phases 2 - 4 and determine information to be provided by the Customer during these Phases.

ARTICLE III - Phase 2 1/8" Scale Sketch Plan

3.1 The purposes of Phase 2 are to produce a 1/8" scale sketch plan including floor plans and front elevation, to develop a preliminary site plan, to complete a client contact work sheet detailing materials and quality level to be included in the house and to prepare a preliminary cost estimate.

3.2 Sketch Plan - The sketch plan will be a 1/8" = 1' scale drawing of the basement, first and second floor plans and front elevation. The floor plans will show room sizes and locations, exterior windows and doors; closet, stair/hallway circulation patterns, kitchen and bathroom layout. The front elevation will show roof style, exterior siding material and window style. A typical example of a sketch plan is attached as Appendix A.

3.3 Site Plan - The preliminary site plan analysis will include determination of basement walkout feasibility, solar orientation, garage and driveway location, zoning and other setback requirements, analysis of covenants restrictions, availability of utilities and privacy considerations. The Owner shall supply information on record plat, topography and covenants.

3.4 Client Contact Worksheet - The worksheet will be filled out and discussed with the Owner. Information from the worksheet will be used in determining the preliminary cost estimate. The Client Contact Worksheet is attached as Appendix B.

3.5 Preliminary Cost Estimate - Based on the sketch plans, site plan and client contact worksheet, the Contractor shall prepare a preliminary cost estimate. The estimate will show cost categories including overhead and profit.

3.6 Review and Redraw - After completion of the above described products the Customer and Contractor shall meet to review the products. Based on this review and a decision by the Owner to continue to Phase 3, the Contractor will do a one time revision of the products which will then become the basis for work in Phase 3.

ARTICLE IV - Phase 3 Preliminary 1/4" Scale Drawings

4.1 The Contractor shall produce the following items during this Phase:

- 1/4" = 1' scale drawings for basement, first and second floor plans

- 1/4" = 1' scale drawings for four (4) elevations

- Preliminary framing and roof plans required to finalize the Contract Estimate

- Preliminary wall section

- A draft contract, including the proposed Contract Amount

- Draft specifications and allowance schedule

4.2 Review and Redraw - After completion of the above described products, the Customer and the Contractor shall meet to review the products. Based on this review and a decision by the Owner to continue to Phase 4, the Contractor will complete a one time revision of the products which will become the basis for work in Phase 4.

4.3   The products produced in Phase 3 shall be of sufficient detail so that the Customer can obtain a construction and permanent financing commitment from a lending institution.

ARTICLE V - Phase 4   Final 1/4" Scale Drawings

5.1   The Contractor shall produce the following items during this Phase:

- Final 1/4" = 1' scale drawings of floor plans and elevations which are fully dimensioned

- Structural information including footing layout, beam and column sizes, framing plans with material size, spacing and grade, detailed wall sections and roof structural plan

- Final kitchen and bath layouts including cabinet and appliance information

- Electrical plan including service location, lighting layout and switch and plug layout.

- Mechanical layout including location of HVAC compressors and air handles, water heaters and pump tanks.

5.2   The Contractor shall prepare and submit all documents required for a building permit. The Owner shall pay for all filing fees which shall be credited to the Owner if the Contractor builds the house.

ARTICLE VI - <u>Payment Schedule</u>

6.1   A fifty percent (50%) deposit is due at the beginning of each Phase for the work to be completed in that Phase. The remaining fifty percent (50%) shall be due at the completion of each Phase.

6.2   The Owner may terminate this Contract at any time and upon such termination shall pay the Contractor in full for any Phase which has been started at the time of termination.

6.3   Schedule of Payment

Phase 1 Product Definition   No Charge

      Phase 2 1/8" Scale Sketch Plan $_____

      Phase 3 Preliminary 1/4" Scale Drawings $_____

      Phase 4 Final 1/4" Scale Drawings $_____

6.4   Additional Services - If the Owner authorizes the Contractor to perform additional design work not covered in this Contract then the following schedule of charges will be used

ABC Builder – Principal	$_____/hr.
Architect	$_____/hr.
Draftsman	$_____/hr.
Secretary	$_____/hr.
Survey Crew (2 Man)	$_____/hr.

ARTICLE VII - <u>Product Ownership and Liability</u>

7.1   The Plans and Specifications have been developed by ABC Builders for its sole use and are copyrighted. The Owner absolves the Contractor of any responsibility for the completeness or accuracy of these plans if the house is not built by the Contractor. If the Owner chooses not to build the house with the Contractor but wishes to obtain originals of the Plans and Specifications, they may do so at an additional cost of $_____.

This Contract shall constitute the sole and entire Agreement between the parties whose signatures are affixed below.

_____   _____
Owner   Date

_____   _____
Owner   Date

_____   _____
ABC Builders   Date

Design Contract

Appendix "A"

SAMPLE 1/8" SCALE SKETCH PLAN

> **Elevation**
> **First Floor Plan**
> **Second Floor Plan**

(NOTE: Builder to insert sketch plans here)

ABC BUILDERS
123 Main Street
Hometown, MD 21234
(301) 231-1234

CLIENT CONTACT WORKSHEET

Page 1 of 4

Name:_____

Address:_____

Telephone: _____   (w) his_____ fax_____
                                        (w) hers_____ fax_____

1.   Approximate starting date:

_____

                                                          YES   NO

2.   Is lot surveyed?   _____   _____
       Grading plan required?   _____   _____
       Permit relocation required?   _____   _____
       Any covenants?   _____   _____

3.   Is lot wooded?   If so, how much?

_____

4.   Septic _____ or Sewer _____
       Septic - disposal?   YES _____   NO _____

5.   Well _____or Water _____

6.   Any demolition?   YES _____   NO _____

7.   WALLS: 2x4_____ 2x6_____ or Other _____
       FLOOR JOIST: 2x10_____ 2x12_____
TJ1_____

8.   Trim description: Traditional_____ Contemporary _____

9.   Main Stairs: Oak _____ Pine _____ Stain _____ Carpet _____
       2ND Stairs: Oak _____ Pine _____ Stain _____ Carpet _____
       Basmt Stairs: Oak _____ Pine _____ Stain _____ Carpet _____

10.   Int. Doors: Paint _____ Stain _____ Flush _____ 6 panel _____

171

11. Ext. Doors: Front - Steel_____ Wood_____ Other_____
    Bsmt. - SGD_____ French _____ Prado_____ Atrium_____
    Other - Steel_____ Wood_____ Other_____

12. Windows: Anderson _____ Pella _____ Alum._____ Other _____

13. Roof: STD_____ Horizon_____ Timberline_____ Wood_____

14. Plumbing: Water Heaters/Size: One (1) _____ or Two (2) _____
    Toilets: Standard_____ Special _____
    Sinks: Standard_____ Special _____
    Whirlpool: YES _____ NO _____
    Description: _____

15. HVAC Specialties:
    Number units _____ heat pump _____ oil _____ other _____
    Comments:_____

	YES	NO
Humidifiers:		
Electronic Filters:		
Laundry Chute:		

16. Electrical: Number recessed: _____
    Estimated Service: 200 AMP_____ 300 AMP_____ 400 AMP_____
    Grounds Lighting: YES _____ NO _____
    Special Requests/requirements:_____
17. Insulation: Walls: R-13 _____ R-19 _____ Other _____
    Ceiling: R-30 _____ R-38 _____ Other _____
    Crawl: _____
    Basement: Walls - _____ Ceiling - _____

18. Drywall: Garage; common wall & clg only: _____
    entire garage: _____

19. Paint/Stain: Unusual requirements - _____

20. Siding: Unusual requirements - _____

21. Finish Driveway: Approx. length - _____
    Stone _____ Asphalt_____

22. Culvert/Apron: Culvert required?   YES _____ NO _____
    Apron required?   YES _____ NO _____

23. Garage Doors: Single Doors _____ Double Doors_____
    Materials: _____

Style: Flush - _____ Paneled - _____

Number of openers - _____

24.   Fireplaces: Raised Flush Hearth Profile Gas
Location Type Hearth Hearth Material Material Fan Doors Logs

_____

_____

_____

_____

_____

_____

25.   Countertops: Kitchen - Laminate_____ Tile_____

          Corian _____ Other _____

        Baths - Laminate _____ Tile_____

          Cultured Marble_____ Other _____

26.   Shower Door: Quantity - Standard _____ Special _____

27.   Appliances: Kitchen: (check if included)

_____ Refrigerator _____ _____ _____ _____

                    19/22cf   24cf   26/27cf   subzero (1or2)

_____ Range - single oven _____ double oven _____

_____ Cooktop - downdraft 4B _____ 6B _____ Other _____

_____ Wall oven - single _____ double _____

_____ Microwave - _____

_____ Disposal - above average _____ best _____

_____ Dishwasher - above average _____ best _____

_____ Washer - _____

_____ Dryer - _____

        Appliances Other: _____

        _____

        _____

        _____

28.   Carpet: Dollar value of allowance (in place) - $/yd _____

29.   Hardwood: Locations _____ _____

        _____ _____ _____

        Strip oak _____ Other _____

30.   Vinyl: Locations:_____ _____

        Dollar value of allowance (in place) - $140/yd. _____

31.   Special items _____

        _____

_____
_____
_____
_____
_____
_____
_____

32. Ceramic Tile/Marble:
Kitchen: _____

Foyer: _____

Master Bath: _____

Other: _____

Other: _____

Other: _____

Other: _____

Other: _____

33. Intercom: YES _____   NO _____

34. Decks: Approx. Sq. Ft. _____
    Special requirements: _____
    _____
    _____

    Porches: Approx. Sq. Ft. _____ Incl. in Contract? _____
    Special requirements: _____
    _____

SPECIAL FEATURES:	YES	NO
Security System	_____	_____
Wall Hung Ironing Board	_____	_____
Central Vacuum System	_____	_____
Greenhouse	_____	_____
Entry Piers, Walls, Gates	_____	_____
Safe(s)	_____	_____
Cedar Closet(s)	_____	_____

This questionnaire has been completed by:

_____   _____
Owner   Date

_____   _____
Owner   Date

_____   _____
ABC Builders   Date

Policy for Owners Using Their
Own Subcontractors and Suppliers

The Policy for Owners Using Their Own Subcontractors or Suppliers defines the "rules of the road" to your client. The policy allows the owners to use their own subcontractors or suppliers but places limitations on scheduling, payments, responsibility and liability so that the builder can maintain control of the overall program.

ABC BUILDERS
123 Main Street
Hometown, MD 21234
(301) 231-1234

CONTRACT PROVISIONS -

In accordance with Article 11 of the Contract between ABC Builders and the Owner, the Owner has the right to perform work related to the project with his own forces and to award separate contracts in connection with other portions of the project or other work on the site.

If the Contractor claims that delay or additional cost is involved because of such action by the Owner, he shall make such claim by Change Order as provide in the Contract. Any costs caused by defective or ill-timed work shall be borne by the party responsible.

The Contractor shall afford the Owner and separate Contractors reasonable opportunity for the introduction and storage of their materials and equipment and the execution of their work, and shall connect and coordinate his work with theirs as required by the Contract Documents.

POLICY FOR SUBCONTRACTORS--

The Owner may hire his own subcontractors to perform work on the project under the following conditions. The Owner shall be responsible for:

- Entering into a subcontract with the Subcontractor

- Coordinating the scheduling and work of the Subcontractor

- Obtaining all permits and inspections required for the Subcontractor's work

- Providing all materials required by the Subcontractor

- Obtaining workmen's compensation and liability insurance for the Subcontractor

- The Owner shall waive the right to hold the Contractor liable for any damage or claims by the Subcontractor against the Contractor

- The Contractor shall specify the amount of credit to be given to the Owner for using his own Subcontractor. This amount shall be deducted from the total contract amount. The Draw Schedule shall be adjusted to reflect reduced payments from the Owner to the Contractor. For example, if the Owner hires his own electrician the Draw Schedule shall be reduced for the mechanical rough draw and the final draw.

- The Owner shall pay the Subcontractor directly and be responsible for obtaining any Waiver of Lien documents required by the lending institution.

POLICY FOR SUPPLIERS-

ABC Builders has provided a list of suppliers that the Owner can choose from when selecting allowance items such as kitchen cabinets, ceramic tile, light fixtures and appliances. If the Owner chooses to use any other supplier, then the Owner shall be responsible for:

- Ordering materials from the supplier and coordinating the delivery of the materials to the job upon seven (7) days' notice by the Contractor.

- Replacing any damaged or defective material including pick-up at no cost to the Contractor.

- Assuming all warranty and service responsibility for the material provided by the supplier.

- Payment - The Owner shall pay all deposits, delivery charges, COD amounts and final payments for all material provided by the supplier. Credits for these payments will be issued by Change Order and shall be deducted from the appropriate draw schedule amount or from the final accounting of the allowance items.

These policies have been agreed to by the Owner and ABC Builders on this date

_____   _____
Owner   Date

_____   _____
Owner   Date

_____   _____
ABC Builders   Date

ABC BUILDERS   **Regulatory Checklist**
123 Main Street
Hometown, MD 21234
(301) 231-1234

Job Name:_____

	Date Submitted	Date Approved	Comments
1. Final Plans with site plan			
2. Permits Architect. Committee Well Building Application Septic			
3. Wall Check/Loc. Survey Final Bldg. Survey			
4. Map to Site			
5. Builders Risk Insur.			
6. Termite Protection			
7. Request for Elec. Svc.			
8. Inspections: County Footings Waterproof Draintile Slab Electrical rough in Plumbing/Mech rough-in Framing Insulation Framing Fireplace Throat Backfill Septic (if applicable) Well (if applicable) Water Test Final Electrical Final Plumbing/Mech. Final Building			

ABC BUILDERS
123 Main Street
Hometown, MD 21234
(301) 231-1234

**Request for Architectural
Committee Approval**

Sample

Mr. Sam Jones, President
Eastside Gardens Architectural Committee
001 Garden Way
Saturn, NY 34543

Dear Mr. Jones:

ABC Builders has contracted with Mr. and Mrs. Tom Williams to build a custom residence on Lot 2, Eastside Gardens, 004 Garden Way. We plan to start construction on March 18, 1998.

Please consider this letter and its enclosures as formal application for architectural approval by the Eastside Gardens Architectural Committee. We look forward to your review and request approval not later than March 1, 1998.

Sincerely,

Brandon Cooke
President, ABC Builders

Enclosures:

General Information Worksheet	( ) Yes	( ) No
Exterior Materials Worksheet	( ) Yes	( ) No
Floor Plans and Elevations	( ) Yes	( ) No
Landscape Plan	( ) Yes	( ) No
Site Plan	( ) Yes	( ) No

ABC BUILDERS
123 Main Street
Hometown, MD 21234
(301) 231-1234

Size

Number of stories _____
Walkout basement   ( ) Yes   ( ) No
Square footage of finished area _____ sq. ft.
    (excluding finished basement)
Square footage of finished basement _____ sq. ft.

Garage

Carport   ( ) Yes   ( ) No   No. of cars _____
Attached Garage   ( ) Yes   ( ) No   No. of cars _____
Garage doors face street   ( ) Yes   ( ) No

Style

Colonial ( )   Contemporary ( )   Tudor ( )   Transitional ( )
Other ( ) Describe:_____
_____

Approval Requested for Other Items

_____ Swimming Pool
_____ Tennis Court
_____ Fencing
_____ Screen Porches
_____ Decks
_____ Outbuildings describe _____
_____ Satellite Dish
_____ Other describe _____

ABC BUILDERS
123 Main Street
Hometown, MD 21234
(301) 231-1234

Roof

Material _____

Color _____

Manufacturer/Selection _____

Sample attached   ( ) Yes   ( ) No

Exterior Siding

Brick

Color _____

Name _____

Mortar Color _____

Sample attached   ( ) Yes   ( ) No

Siding

Type_____

Color _____

Name _____

Sample attached   ( ) Yes   ( ) No

Exterior Paint Colors

Trim_____

Siding _____

Doors _____

Windows _____

If prefinished - manufacturer;_____ color;_____

Gutters and Downspouts

Material _____

Color _____

ABC BUILDERS
123 Main Street
Hometown, MD 21234
(301) 231-1234

Application for Utility Service

Location
    Lot _____   Block _____   Subdivision _____
    Address   _____
                   _____
                   _____

Type of Services Requested

    Electric ( )   Gas   ( )   Telephone ( )
    Cable TV ( )   Water ( )   Sewer   ( )

    Site plan attached with meter locations   ( ) Yes   ( ) No

    Date service required by _____

Service Load Information
    Electric
            Service panel _____ amps
            Total resistance heat load _____ kw
            Other loads: Electric Dryer ( )   Range ( )   Wall oven ( )
                    Other ( ) Describe _____
            Air conditioning load _____ tons

    Gas
            Line size requested _____ inches
            Total heat load _____ BTU/HR
            Other gas appliances: Dryer ( ) Range ( )   Wall oven ( )
                    Fireplace ( )

    Sewer
            Number of bedrooms _____
            Number of bathrooms_____

    Water
            Line size requested_____ inches

ABC BUILDERS   **Building Permit Checklist**
123 Main Street
Hometown, MD 21234
(301) 231-1234

Plans

_____1/4" scale drawings
_____foundation/footing plans
_____cross section
_____roof plan with structural information
_____structural plans

Site Plan

_____well location
_____septic field
_____topography
_____driveway and parking pad
_____building restriction lines
_____easements and setbacks

Application

_____Lot information
_____Lot and block
_____Subdivision
_____Street address
_____Tax I.D. number
_____Zoning
_____Acreage or square footage

_____Owner information
_____Name
_____Current address
_____Telephone number

_____Special requirements
_____Number of bedrooms
_____Number of full and half baths
_____Square footage
_____Height
_____Energy calculations

_____Bonds and fees
_____Impact fee calculations

_____Bond requirements including forms or letters of credit
_____Application fee calculations

Sediment and Erosion Control/Storm Water Management

_____Disturbed area
_____Impervious area-runoff calculation
_____Grading plan
_____Sediment and erosion control plan
_____Drainage pipe sizes -- including driveway

CLIENT INFORMATION PACKAGE

Client Package

How you communicate with your clients sets the stage for how smoothly and profitably the project will go. The Client Information Package is a set of documents designed to create realistic expectations on the part of the buyer and, at the same time, to establish confidence and credibility in the builder. This information package should deal with all the responsibilities of the clients. In addition, the forms will explain the scheduling and construction process, discuss quality issues and detail the change-order procedure. The following documents need to be included:

Scheduling - A general outline of the milestones in construction over the course of the contract should be denoted for the client. Additionally, a paragraph should explain why scheduling variations occur, stressing such items as weather, subcontractor reliability, and conflicting schedules. You must also mention that periodically things happen which are not predictable - inspections, late material deliveries, or a brief wait while someone becomes available.

Client Selections - It should be clear to the client that numerous selections must be made during the course of construction. It is imperative to state that late selections may delay the move-in date and to stress again the responsibility of the client in the selection process. A detailed schedule of items that need to be selected at a particular time should be included. It is best to meet with the clients and complete the list of items that need to be selected at that time. Additionally, a form should be provided for recording the client's selections. Clients should sign all selections to ensure that there will be no problems later on.

Contact List - A list of all suppliers from whom you wish the clients to select and purchase materials. If possible, include a specific contact person.

Change Orders - The change-order procedure is a complicated and important one and a detailed explanation of change orders is included with the Client Information Package.

Quality Control and Material Characteristics - A brief explanation of key materials and their characteristics. This will alleviate undue concern by the clients when cracking, shrinkage or color variations occur. The most common items to be addressed include brick and mortar, ceramic tile and marble, sheet products such as drywall and concrete work.

Standard Prices - A detailed listing of your most commonly requested extras and their prices. This helps minimize surprise and prepares your client for the "reality" of pricing extras. You may want to have clients sign this form to lock in the optional prices shown with the contract. Note: This listing is included in the Contract Documents section. You need to decide where it is most appropriate.

Client Notebook

Building a custom home is a complicated process. You are involved with the client for a long period of time and there is a tremendous amount of information that must be collected, stored and accessed in an efficient manner.

One practical method of organizing and managing the custom homebuilding process is to have a client notebook for each customer. The notebook will be divided into eight sections. These sections will be the same for each client and information will be filed in a consistent manner. This will enable any member of the builder's organization to find specific information quickly. The eight sections are:

- Permits
- Contract/Specification
- Costs/Changes
- Income/Draws
- Client Selections/Details
- Subcontractors
- Orders
- Correspondence

The following Client Notebook Content sheets show the type of information that should be completed in each section of the Client Notebook.

BUSINESS BUILDER:
NOTHING impresses clients more than your picking up their notebook and immediately finding the information you need. Nothing looks worse than, "Sorry, I can't seem to find it."

ABC BUILDERS
123 Main Street
Hometown, MD 21234
(301) 231-1234

Realtor/Client
Registration Form

One situation to be avoided at all costs is a dispute as to the source of a sale. Disagreements as to who is owned a commission does not sit well in the real estate community, and can lead to disgruntled customers and a lack of future referrals. The key to eliminating these troublesome situations is the use of a referral log and the realtor registration form. The system works as follows:

For Custom Builders
When a realtor refers a client, the realtor's name and contact information, as well as the client information, is logged and the date recorded. At the time the client's name is checked against the log to verify that the buyer has not been previously referred. That's why an electronic spreadsheet or database is best – you can do a search by name and date and the information is immediately available.) If I find that the buyer has been previously referred, you inform the new referring agent of the date and name of the original referral. If no previous referral exists, you give one copy of the Realtor Referral card to the referring party, and the other copy is filed and logged.

The key elements necessary for a good Realtor Registration form are:

Name, address, phone number of referring party.
Name, address, phone number of buyer.
Date of referral.
Rules and policies concerned with referral.
Amount of referral (typically a percentage of the sales price.
Period of time referral is in effect (usually six months or a year.)
Any special conditions or caveats. (Only one referral per client).

For Production Builders
Conceptually the referral policy for production builders is similar that those of custom builders with two exceptions:
The referral logs and forms are kept and distributed on a project by project basis.
To receive a referral the buyer must physically visit the site. Typically the referring party is required to accompany the buyer; this is frequently waived with an advance phone call.

Realtor/Client Registration Form   Date _____

Name of Client _____
Address _____
City_____ State _____ Zip _____
Phone Number _____ Fax _____
Name of Realtor _____
This registration is valid for six months from date of registration. Client must accompany realtor at time of registration.

ABC BUILDERS
123 Main Street
Hometown, MD 21234
(301) 231-1234

Request for Information

Request For Information
Whenever an interested party contacts your firm and requests information, this data should be collected and logged for future follow-up. The following items are critical on any form of this type:
1. Name, address, home and work phone number of the party.
2. Information requested.
3. Date of the request (and date information was sent).
4. How did the party find out about you. These should be broad categories, rather than trying to get to specific. Don't waste their time, and lose their interest.
5. Is the party working with a real estate agent?
6. Level of interest (Need immediately or sometime soon, vs. just looking.)
7. Any specific requests or preferences. For example. Only interested in the north side of town. Need 4 bedrooms or 3 car garage, etc.

A standard letter should accompany the material, and a follow-up procedure should be in place. A phone call 3-5 days after the materials are sent (to confirm their arrival) is ideal.

Request for Information   Date _____

Name _____

Mailing Address _____

City/State/Zip _____

Phone Number _____ Fax _____

Information requested _____

Brochures: ☐ 4 page positioning   ☐ 2 page flyer

                     Model Home Information (model) _____

                     Development Information (subdivision) _____

                     ☐ Other _____

Sent (Date) _____ By _____

Follow up: Received _____ Verified by _____

Urgency of need: ☐ Immediate ☐ 3-6 mo.   ☐ 1 year   ☐ Just looking

Comments: _____
_____
_____
_____

ABC BUILDERS   **Prospect Card**
123 Main Street
Hometown, MD 21234
(301) 231-1234

Date _____

Prospect Name _____

Mailing Address _____

City/State/Zip _____

Phone Number _____ Fax _____

How did you hear about (Your Company)? ☐ Friend   ☐ Newspaper Ad   ☐ Drove by

☐ Other _____

How quickly do you need a new home? ☐ NOW   ☐ Within 6 months ☐ Within 1 year☐

☐ Just looking, thanks.

What made you interested in (Your Company)?   ☐ Value   ☐ Quality   ☐ Price   ☐ Location
☐ Reputation   ☐ Home Design ☐ Other _____

What did you like best about the home itself? ☐ Exterior design ☐ Interior layout
☐ Spaciousness   ☐ Indoor/outdoor living ☐ Storage ☐ Quality of finishes/components
Specific comments: _____

_____

_____

Follow-up
Date _____   Action _____

Date _____   Action _____

Date _____   Action _____

ABC BUILDERS
123 Main Street
Hometown, MD 21234
(301) 231-1234

Referral Card

One of your best sources for new clients is referrals from existing or previous clients. However, in order to get those referrals they have to be satisfied customers, and you usually have to ASK for the referral.

The referral card is something that the builder or customer service manager will usually fill out. You may, however, wish to include a copy in the customer service package or in periodic mailings to past clients.

Make sure that you not only get the name and address and phone number of the referral, but also their relationship to the person giving the referral. And you may want to treat referrals in a special manner. Invite them to a model home after hours or meet them in the home of the referring couple. Present them with a special gift that is related to the home. To set everyone at ease, it should not be simply a sales call.

Date _____

Name of Referral _____

Current Address _____

City/State/Zip _____

Phone Number _____ Fax _____

Name of Person Giving Referral _____

Current Address _____

City/State/Zip _____

Phone Number _____ Fax _____

Relationship of referral to person giving referral ☐ Friend ☐ Co-worker ☐ Relative
☐ Aquaintance ☐ Other Specifics _____

Follow-up
Date _____ Action _____

Date _____ Action _____

Date _____ Action _____

ABC BUILDERS
123 Main Street
Hometown, MD 21234
(301) 231-1234

SAMPLE

Dear Client:

Thank you so much for choosing us to build your new home. We at ABC BUILDERS believe that the keys to a good relationship are realistic expectations and a conscientious effort by the parties to communicate effectively. In preparation for our working together for the next six to nine months, we have put together this handout to assist in answering some of the many questions which will arise during the planning and construction of your new home. The material is divided into four main sections including scheduling, client selections, change orders, and quality control/material characteristics. Read this material carefully, and if you have any questions please feel free to call our office.

It is our pleasure to have the opportunity to work with you in building a home that will be a source of pride for all of us.

Sincerely,

Owner
ABC BUILDERS

ABC Builders
123 Main Street
Hometown, MD 21234
(301) 231-1234

Due to normal variations in the custom construction process including the ordering, manufacture and installation of your selections, the completion date may be forty-five days earlier or later than projected. Your final move-in date will be determined approximately thirty days before the scheduled completion date.

While every effort will be made to accommodate you, there are the day-to-day uncertainties of conflicting schedules and subcontractor reliability. Unlike the routine steady pace of the manufacturing business, you will find the progress on your house to come in spurts, with periods of great activity and other times when it appears that nothing is happening. Often there are things happening which are not readily visible such as inspections, deliveries, or just a one or two day wait for someone to be available for a certain task. Historically our deliveries have been very close to schedule. As your home nears completion we will keep you informed as to the final delivery date.

ESTIMATED CONSTRUCTION MILESTONES
(cumulative in weeks)

	7-month contract	8-month contract	9-month contract
First Deck	3	4	5
Roof Sheathed	7	9	11
Mechanicals Complete	11	13	6
Drywall	14	16	20
Trim	20	22	27
Final	28	32	36

Client Selections

There are many selections to be made prior to and during the construction of your home. We have attached a "Client Selection Schedule" detailing what decisions will be made at each meeting. Our staff is here to advise you in this selection process. When a selection is needed, we ask that a decision be made as quickly as possible. Certain items may take several weeks to arrive. For example, ceramic tile may take 4 - 6 weeks and light fixtures often take 6 - 10 weeks.

We have attached a list of approved sources to contact. We suggest that you visit these retailers when making your selections. If you wish to make a selection elsewhere we are generally amicable but we must first approve the source.

Again, we would like to stress the importance of making timely selections. A LATE SELECTION MAY DELAY YOUR MOVE-IN DATE.

ABC BUILDERS   **Client Selection Schedule**
123 Main Street
Hometown, MD 21234
(301) 231-1234

Client Name _____

MEETING No. 1 - FRAMING PHASE:
INITIAL SELECTIONS - PRIOR TO FOOTINGS
(usually within 10 days of signing contract)
• Siding Material: Options being, Vinyl, Aluminum, Brick, Wood/Cedar Roofing Material
• Location of: ceramic, hardwood, carpet, vinyl

NOTE: Client advised at this time to visit cabinet manufacturers to come to a decision
on cabinet type/color by third meeting.

MEETING No. 2 - ROUGH-IN PHASE:
(within several days after first meeting)
• Exterior Door Style
• Interior door sizes/Pocket door locations
• Window details
• Final decision on kitchen layout
(Layout only, not style of cabinets, colors, etc. This will come at a later date)
• Countertop material to be used - Options: Formica/marble/etc.
• Appliance types - location of:
• Final decision on bathroom layout (layout only)
• Countertop material for bath's - cultured marble/Formica/etc.
• Final bathtub selection - shower/tub valve selection
• Finalize placement of medicine cabinets/mirror locations in baths.
• Finalize electrical layout within house - notification of any special pre-wire (electricity,
telephone, etc.).

NOTE: Client is advised to visit light fixture retailers to finalize fixture order. Client is
also advised to visit ceramic tile dealers to finalize tile order at our next meeting.

MEETING No. 3
SECONDARY SELECTIONS
(Within 6 to 8 weeks from previous meeting)
• Siding; trim colors/shutter color/garage door color
• Interior stain/paint colors (hardwood flooring /trim/ handrails/ walls/ ceilings, etc.)
• Exterior door hardware selection
• Interior door hardware selection

• Final decision on kitchen/bath vanity style/color, etc.
• Countertops (finalize color)
• Ceramic tile
•Final plumbing fixture decisions: toilets/sinks/showers/shower doors/faucets/towel bars/ mirrors/medicine cabinets
• Verify light fixture order has been placed. (Need approx. 8 wks. lead time for delivery)

NOTE: It is now suggested Client begin thinking of appliance decisions, carpet selections, vinyl flooring selections.

MEETING No. 4
FINAL SELECTIONS
*	Final flooring decisions: vinyl, carpet
*	Final appliance decisions
*	Final bath hardware ordering (mirrors/ medicine cabinets/specialty items)

ABC BUILDERS
123 Main Street
Hometown, MD 21234
(301) 231-1234

Client Name _____

1. ROOFING
Shingles: _____
Flashing: _____

2. SIDING
Material: _____
Color: _____

3. DOOR SELECTIONS

	Location	Selection

Exterior:

_____
_____
_____
_____

Garage Door: _____
With opener? _____

Interior: _____
_____
_____
_____
_____
_____
_____
_____

4. WINDOW SELECTION

_____
_____
_____
_____
_____

_____
_____
_____

5.   MASONRY

Exterior: _____ mortar: _____
Interior (fireplace): _____ mortar: _____
Other: _____ mortar: _____

6.   PAINTS/STAINS

| | Location | Selection |

Exterior:

_____
_____
_____
_____
_____
_____

Interior: _____
_____
_____
_____
_____
_____

7.   PLUMBING FIXTURES

Powder Room - Tub/Shower: _____
        Toilet: _____
        Sink(s): _____
        Faucets: _____
        Hardware _____

Master Bath - Tub/Shower: _____
        Toilet: _____
        Sink(s): _____
        Faucets: _____
        Hardware: _____

Other Bath (location: _____)
    Tub/Shower: _____
    Toilet: _____
    Sink(s): _____
    Faucets: _____
    Hardware: _____

Other Bath (location: _____)
    Tub/Shower: _____
    Toilet: _____
    Sink(s): _____
    Faucets: _____
    Hardware: _____

8.   CERAMIC TILE (list tile color/grout color/sq. ft.)
Powder Room: Walls: _____
    Floors: _____

Master Bath: Bath Walls: _____
    Shower Walls: _____
    Bath Floor: _____
    Shower Floor: _____

Other Bath: (location: _____)
    Bath Walls: _____
    Shower Walls: _____
    Bath Floor: _____
    Shower Floor: _____

Other Bath: (location: _____)
    Bath Walls: _____
    Shower Walls: _____
    Bath Floor: _____
    Shower Floor: _____

Kitchen: Walls: _____
    Deco tiles: _____
    Floor: _____

Other areas to receive ceramic: _____
_____
_____
_____

9.   CABINETS/VANITIES:

	Manufacturer	Style/Color	Hardware
Kitchen:	_____	_____	_____
	_____	_____	_____
Laundry:	_____	_____	_____
	_____	_____	_____
Pwdr. Rm:	_____	_____	_____
	_____	_____	_____
Master Bath:	_____	_____	_____
	_____	_____	_____
Other Bath: (Location)	_____	_____	_____
	_____	_____	_____
Other Bath: (Location)	_____	_____	_____
	_____	_____	_____

10.   COUNTERTOPS (list manufacturer/color no./size)

Kitchen: _____
_____

Laundry Room: _____
_____

Powder Room: _____
_____

Master Bath: _____
_____

Other Bath (Location): _____
_____

Other Bath (Location): _____
_____

11.   LIGHT FIXTURES & LOW VOLTAGE (If nothing listed check under "orders")

Location	Selection(s)
Door bell chimes	_____
Intercom & speakers	_____
Exterior	_____

Foyer _____

Dining Room _____

Master Bedroom _____

Bedroom #2 _____

Bedroom #3 _____

Bedroom #4 _____

Walk-in Closets _____

Laundry Room _____

First Floor Hallway _____

Second Floor Hallway _____

Master Bath _____

Other Bath #1 _____

Other Bath #2 _____

Powder Room _____

Kitchen _____

Great Room _____

Garage _____

Basement _____

Other _____

Other _____

Other _____

12.   APPLIANCES
Stove/Oven/Microwave: _____
_____

Refrigerator: _____
_____

Dishwasher: _____
_____

Disposal: _____
_____

Washer: _____
_____

Dryer: _____
_____

Central vacuum: _____
_____

Other: _____
_____

13.   FLOORING (Vinyl/Carpet/Hardwood - please specify)
Location   MFR/Color/Size

_____   _____
_____   _____
_____   _____
_____   _____
_____   _____
_____   _____
_____   _____
_____   _____
_____   _____
_____   _____

14. MIRRORS/MEDICINE CABINETS/SHOWER DOORS:

Location   Selection

_____   _____
_____   _____
_____   _____
_____   _____
_____   _____
_____   _____
_____   _____
_____   _____
_____   _____
_____   _____
_____   _____
_____   _____

15. MISCELLANEOUS SELECTIONS/EXTRAS

_____
_____
_____
_____
_____
_____
_____
_____
_____
_____
_____
_____
_____
_____
_____
_____
_____

ABC BUILDERS
123 Main Street
Hometown, MD 21234
(301) 231-1234

BRICK SUPPLIERS

Supplier #1
Contact
Address
Telephone

Supplier #2
Contact
Address
Telephone

Supplier #3
Contact
Address
Telephone

Supplier #4
Contact
Address
Telephone

CABINETS AND VANITIES

Supplier #1
Contact
Address
Telephone

Supplier #2
Contact
Address
Telephone

Supplier #3
Contact
Address
Telephone

Supplier #4
Contact
Address
Telephone

LIGHT FIXTURES

Supplier #1
Contact
Address
Telephone

Supplier #2
Contact
Address
Telephone

Supplier #3
Contact
Address
Telephone

Supplier #4
Contact
Address
Telephone

CERAMIC TILE

Supplier #1
Contact
Address
Telephone

Supplier #2
Contact
Address
Telephone

Supplier #3
Contact
Address
Telephone

Supplier #4
Contact
Address
Telephone

FLOOR COVERINGS

Supplier #1
Contact
Address
Telephone

Supplier #2
Contact
Address
Telephone

Supplier #3
Contact
Address
Telephone

Supplier #4
Contact
Address
Telephone

ABC BUILDERS
123 Main Street
Hometown, MD 21234
(301) 231-1234

Lighting Fixture Selections

Date: _____

Client: _____

Location	Type of Fixture	Quantity	Selection/Supplier

Change Orders

You may request changes in the work being done on your home. These requests are referred to as Change Orders and may consist of an addition, a deletion or a modification.

Change Orders cost different amounts depending on when they are received during construction. For example, an added entryway will be much less expensive if ordered during the early framing phase of construction than the same entryway ordered after the drywall has been installed. A price sheet for many standard items and options is attached for your information. These prices assume adequate notice, and may increase for changes requested after related work has been done.

Occasionally, a Change Order may delay your move-in date, due to delays in receiving materials, special ordering of items, or re-coordinating subcontractors to modify work that has already been completed. Additionally, some Change Orders are prohibitively expensive if a substantial amount of work has already been done. Careful review of your plans prior to construction will minimize the need for Change Orders.

When a Change Order is instituted, two Change Order forms will be sent to you. These will include a short description of the work and the corresponding charge or credit. Please sign and return the original and keep the extra copy for your records. A sample change order form is included.

ABC BUILDERS
123 Main Street
Hometown, MD 212342
(301) 231-1234

Change Order #____

Date: _____

Client: _____

This is to authorize ABC BUILDERS to do, or have done, the following described work:

_____

_____

_____

_____

_____

_____

_____

_____

_____

_____

_____

_____

_____

_____

*I understand this work is not included in any previous contract, prior order for extra work, plans or specifications. For the above extra work I agree to pay ABC BUILDERS the sum of _____
Dollars ($)_____ which is in addition to any and all previous contracts and prior orders for extra work. Per the Contract (Article 4.3) the amount of this Change Order is due and payable at the time of our next draw.

*Please sign and return one copy. Retain the other copy for your records.

_____   _____
Owner   Date

_____   _____
Owner   Date

_____   _____
ABC BUILDERS   Date

ABC BUILDERS
123 Main Street
Hometown, MD 21234
(301) 231-1234

**Quality Control &
Material Characteristics**

Brick – Brick is a natural material and thus varies in color and texture. All the brick ordered for your home will be from the same run.

Mortar – Mortar may have slight color variations depending on water content, sand color and drying conditions. The same mortar will be used throughout the project, and all sand will be supplied by the same quarry.

Exterior Wood Stain for Cedar Siding - Due to the nature of semi-transparent stains, color grain and texture, a trial sample is recommended to get the exact color you require.

Interior Wood Stain - Interior wood that is to be stained should be free of knots and blemishes and should be clean prior to staining. Three types of variations occur when staining wood.

1. Species - Typically there are different wood species used in a house. The most common will be white pine, which is used for casing, base, chair rail and other common trim material. In addition, fir may be used for interior or exterior doors. Specialty trim can be made from oak, poplar, cherry, walnut, and mahogany. One stain will look different as it is applied to each type of wood.

2. Coverage - Stains are designed to be wiped off after allowing for a short setting time. If a darker color is desired a darker stain should be used rather than a second application of the lighter stain.

3. Grain - Each individual piece of wood "takes" stain differently depending on the grain. Flat grains absorb less stain and therefore appear lighter than end grains, which absorb more stain. This is particularly noticeable in oak, which has an open, porous end grain. Oak wood floors, even when stained, sealed and polyurethane finished may appear rough because the material is being absorbed into the end grain.

Ceramic Tile - As with other materials, color variations can occur in ceramic tile. For example, a 4x4 "field" tile and a 4x4 "bullnose" tile with the same color number may be slightly different because they are glazed and fired at different times. All tile will be ordered from the same "run" and, in general, a slight over order is made. We suggest that you store the extra tile for future use in case of chipping, broken tile, etc. The same general conditions that apply for mortar apply for ceramic tile grout. ABC BUILDERS cannot be responsible for normal variations in this material.

Marble - Marble is a natural material with variations in color and thickness. Prior to laying the marble a sample from each box will be examined to insure reasonable uniformity and you will be consulted if significant variations are found. Again, ABC BUILDERS cannot be responsible for normal variations in this material.

Drywall - Drywall sheets are joined together at seams and corners with drywall tape and joint compound. The nail holes are covered with joint compound. This system does not result in a completely uniform surface such as is achieved by plaster. Therefore, variations in the final painted surface are expected where the seam, corners and nails are covered by joint compound. ABC's job is to come as close to plaster as can reasonably be expected. Several conditions that are critical need to be addressed. First, semi-gloss paint magnifies any imperfections and is not recommended except for kitchen and bath walls, if requested. Second, any light reflecting at a "flat" angle to the walls or ceiling creates shadow lines at seams and nails. This same condition occurs for cathedral ceiling and outside window light when viewed at flat angles. A stippled or textured ceiling should be considered if these conditions occur over broad areas.

Painting - Matching colors for stock paints can be done for a period of time after the house has been painted and before there are color changes due to fading, wear and dirt/dust buildup. Matching colors for mixed colors cannot reasonably be expected and therefore any leftover paint should be kept for touch-up use during the first year.

Concrete Shrinkage - As the basement and garage slabs "cure" and become hard a certain amount of water evaporates from the concrete. This causes shrinkage cracks to occur. Wire mesh is placed in the concrete to control this cracking but does not totally prevent the cracking. Horizontal cracks up to the thickness of a dime are not uncommon. Vertical displacement of cracks should be small.

Walk-Through Inspection

There are two objectives in a final walk-through inspection with the client.

The first objective is to make a list of items that need to be corrected prior to move-in. This is a somewhat negative task in that you as a builder are dealing with incomplete or unsatisfactory items.

The second objective is to provide information to the new owners about their house and how it works. This is a positive experience for the owners.

The Walk-Through Inspection Information Checklist provided in this section should be used as a guide in fulfilling the second objective of the walk-through process. It lists the key areas of information that should be covered in the walk-through and provides a sample service call list.

BUSINESS BUILDER
If you take the time to do a comprehensive walk-through, explain how everything works and provide names and telephone numbers of contacts for service calls you MIGHT not have to answer that 2 a.m. phone call from your client who can't set the thermostat to emergency heat.

ABC BUILDERS
123 Main Street
Hometown, MD 21234
(301) 231-1234

Exterior

- Location of septic tank and/or sewer. Cleanout septic tank should be pumped out every _____ months.
Name and telephone of septic service:
Name _____
Tel. _____

- Location of hosebibs

- Location of exterior receptacles. Note that these receptacles are on a GFI circuit(s) and if they trip they must be reset at the following GFI location(s) _____ and _____.

- Utilities - Location of meters and contact:
Name _____
Tel. _____

- Electric
Name _____
Tel. _____

- Telephone
Name _____
Tel. _____

- Gas
Name _____
Tel. _____

- Cable TV
Name _____
Tel. _____

- Alarm
Name _____
Tel. _____

- Heating Oil/Propane
Name _____
Tel. _____

- Garbage Collection
Name _____
Tel. _____

- Public Water & Sewer
Name _____
Tel. _____

- 	Keys/Door Openers:

- ( ) Demonstrate garage door openers, emergency release, locking bar and present keys and openers

- ( ) Present two sets of exterior keys and show which doors they operate

Heating and Air Conditioning
- Location of compressors - show disconnect boxes and explain automatic defrost cycle

- Demonstrate operation of typical floor and ceiling registers

- Air handles
    - ( )	Location
    - ( )	Filter changing
    - ( )	Filter sizes _____ _____
    - ( )	Condensate pans and drains show where condensate is discharged and explain about clogging and backups
    - ( )	Gas valves and restart procedure
    - ( )	Oil tank and fill location
    - ( )	Propane tank and fill location
    - ( )	Location of manual and automatic dampers

- Humidifiers:
    - ( )	Water valves
    - ( )	Operating procedures
    - ( )	Drum replacement - brand _____ size _____

- Electronic air cleaner:
    - ( )   Operating procedures
    - ( )   Replacement parts - brand _____ size _____

- Thermostats
    - ( )   Location(s)
    - ( )   Operating procedures
    - ( )   Emergency heat setting
    - ( )   Night set back

- Whole House Fan
    Attic Ventilation Fan
    Bathroom heat/light/vent fans
    Special items _____
    _____
    _____

Plumbing
- Location of main house valve

- Demonstration of typical individual fixture cut-off valves
    - ( )   Location of hose bib valves and demonstrate how to drain in the Fall
    - ( )   Water heaters
    - ( )   Temperature setting
    - ( )   Draining
    - ( )   Pressure relief valve

- Well pump

- Whirlpool tub demonstration

- Pressure Tank
    - ( )   Pressure setting

- Sewage ejector pumps

- Under sink ejector pumps

- Location and operation of all gas valves
    Special items _____
    _____
    _____

Appliances

Demonstrate the operation of each kitchen/laundry appliance including presentation of owners' manual and warranty cards.

- Range/Wall oven
  - ( ) Cleaning
  - ( ) Filters and vents

- JennAire
  - ( ) Filters
  - ( ) Grease jars
  - ( ) Accessories
  - ( ) Changing burners
  - ( ) Vents

- Dishwasher

- Disposal
  - ( ) Jams
  - ( ) Unclogging
  - ( ) Reset

- Instant Hot
  - ( ) Temperature setting

- Refrigerator/Freezer
  - ( ) Ice maker valves
  - ( ) Temperature setting

- Microwaves

- Trash compactor
  - ( ) Bags
  - ( ) Jams

- Washer
  - ( ) Hose connection
  - ( ) Filters

- Dryer
  - ( ) Vents
  - ( ) Gas connection

Electrical
- Location of meter and panel

- Discussion of circuits and circuit breakers

- Labelling of circuits

- Discussion of GIF circuits
    - ( )   How to test and reset breakers
    - ( )   Location of <u>ALL</u> GFI breakers

- Location of all service switches such as at pump for Whirlpool

- Smoke detectors

- Low Voltage

- Intercom
    - ( )   Master control panel
    - ( )   Individual speakers
    - ( )   Features

- Central Vacuum
    - ( )   Main unit filters and cleaning
    - ( )   Demonstration of typical outlet
    - ( )   Explanation of accessories

- Alarm System
    - ( )   Full demonstration by alarm company
    - ( )   Access code

- Telephone(s)

- Television
    - ( )   Location of attic antennae wires
    - ( )   Location of main cable line

ABC Builders
123 Main Street
Hometown, MD 21234
(301) 231-1234

Item	Company	Telephone #
Electrical	_____ _____	Normal_____ Emergency_____
Plumbing	_____ _____	Normal_____ Emergency_____
Heating/AC	_____ _____	Normal_____ Emergency_____
Appliances	_____ _____	Normal_____ Emergency_____
Garage Doors & Openers	_____ _____	Normal_____ Emergency_____

Kitchen and Bathroom
Design Worksheets

Kitchen and bathroom design and implementation can "make you" or "break you" as a custom builder. Many times, these rooms have the highest visual impact of the entire house. At the same time they are the most expensive to build and are the source of many potential mistakes and cost overruns.

Careful planning and organization starting with day one on the job will minimize your kitchen and bathroom problems. The worksheets contained in this section are designed to force you to cover all aspects of kitchen and bathroom design and implementation. They need to be supplemented with cabinet and vanity layout drawings. Sample graph paper layout sheets are provided (only in the hard-copy documents) to assist in these layouts.

The kitchen design worksheets are divided into four sections -cabinets, appliances, countertops and mechanical. When completing the worksheets, be sure to consult with your plumber, electrician, HVAC contractor and cabinet supplier.

The bathroom design worksheets are divided into five sections -- vanities, countertops, accessories; plumbing fixtures; and ceramic tile.

BUSINESS BUILDER:
From preconstruction to occupancy, give the kitchen and baths your maximum attention. Homeowners happy with their kitchen will usually be happy with their house. Homeowners frustrated by their kitchen will usually be unhappy and find other things to be frustrated about.

ABC BUILDERS
123 Main Street
Hometown, MD 21234
(301) 231-1234

Name of Manufacturer_____

Name of Supplier_____
Telephone Number_____

Kitchen layout   ( ) Attached   ( ) Per Plans

Cabinet Style_____

Base Cabinet Doors_____
_____

Wall Cabinet Doors_____
_____

Glass Doors   ( ) Yes   ( ) No
  Glass Supplied by_____

Door Hardware_____

Door Hinges_____

Wood_____

Stain Color_____

Formica_____

Bulkheads   ( ) Yes   ( ) No
  Material_____
  Installed by_____

Panels for Appliances
  ( ) Dishwasher   ( ) Trash compactor
  ( ) Refrigerator   ( ) Freezer
  ( ) Other_____

Refrigerator
        Manufacturer _____
        Model # _____
        Color _____
        Size _____ Width _____
        Height _____ Depth _____
        Ice Maker Line   ( ) Yes   ( ) No
        Door Swing _____
        Wood Door Panels: ( ) Yes   ( ) No

Dishwasher
        Manufacturer _____
        Model # _____
        Color _____
        Wood Door Panels: ( ) Yes   ( ) No

Disposal
        Manufacturer _____
        Model _____
        Location in sink: ( ) Right Bowl   ( ) Left Bowl   ( ) Center Bowl

Trash Compactor
        Manufacturer _____
        Model # _____
        Color _____
        Width (inches) _____
        Wood Door panels: ( ) Yes   ( ) No

Wall Oven
        Manufacturer _____
        Model # _____
        Color/Style _____
        Size: width _____ height _____ depth _____

Microwave
        Manufacturer _____
        Model # _____
        Color _____
        Mounting Kit _____

Range

    Manufacturer _____

    Model # _____

    Size _____ width _____

    Drop-in cut-out size

Range Hood

    Manufacturer _____

    Model # _____

    Venting requirements _____

     and duct size _____

Cook Top

    Manufacturer _____

    Model _____

    Color _____

    Accessories _____

    _____

    Countertop cut-out size _____" x _____"

    Radius in corners _____"

Freezer

    Manufacturer _____

    Model # _____

    Color _____

    Door Swing _____

Instant Hot

    Manufacturer _____

    Model # _____

Washer

    Manufacturer _____

    Model # _____

    Color _____

Dryer

    Manufacturer _____

    Model # _____

    Color _____

    Gas _____ Electric _____

ABC BUILDERS
123 Main Street
Hometown, MD 21234
(301) 231-1234

First - Top Material _____

Manufacturer _____

Color _____

Location _____

Edge Detail _____

Backsplash _____

Second -Top Material _____

       Manufacturer _____

Color _____

Location _____

Edge Detail _____

Backsplash _____

Eating Bars

Overhang _____ inches

Height _____ 30" _____ 36" _____ 40"

Special Items _____

_____
_____
_____
_____
_____
_____
_____
_____
_____

ABC BUILDERS
123 Main Street
Hometown, MD 21234
(301) 231-1234

PLUMBING

Kitchen Sink
    Manufacturer _____
    Model # _____
    Color _____
    Cut-out size or template _____
    Number of holes _____
    Faucet _____
    Number of Bowls   ( ) Single   ( ) Double   ( ) Triple

Salad or bar sink
    Manufacturer _____
    Model # _____
    Color _____
    Cut-out size _____ or template _____
    Faucet _____

ELECTRICAL
Wiring
    Wiring requirements for separate circuits for appliances
    Refrigerator _____ amps
    Freezer _____ amps
    Dishwasher _____ amps
    Disposal _____ amps
    Instant Hot _____ amps
    Microwave _____ amps
    Range _____ amps
    Range Hood _____ amps
    Cooktop _____ amps
    Wall Oven _____ amps
    Washer _____ amps
    Dryer (elec) _____ amps
    Dryer (gas) _____ amps
    Iron _____ amps

Lighting requirements

       Under cabinets _____

       Recessed lights in bulkhead or ceiling _____

       Island lights _____

VENTILATION

       Toe kick heat vents   ( ) Yes   ( ) No

       Cooktop vent _____ size

       Power unit on outside   ( ) Yes   ( ) No

       Wall oven vent   ( ) Yes   ( ) No _____ size

       Range hood vent   ( ) Yes   ( ) No _____ size

ABC BUILDERS
123 Main Street
Hometown, MD 21234
(301) 231-1234

**Bathroom Design Worksheet
Vanities, Countertops, Accessories**

Bathroom # _____   Location _____

VANITIES

Manufacturer _____
Supplier _____
Vanity layout   ( ) attached   ( ) per plans
Vanity style _____
Wood _____
Stain color _____
Formica _____
Door hardware _____
Door hinges _____

COUNTERTOPS

Material _____
Manufacturer _____
Color _____
Edge detail _____
Backsplash _____
Special Items _____

_____

ACCESSORIES

Medicine chest _____
Towel bars _____
Towel rings _____
Toilet paper _____
Shower/tub enclosure _____
Other _____

_____
_____

ABC BUILDERS
123 Main Street
Hometown, MD 21234
(301) 231-1234

Bathroom Design Worksheet
Plumbing Fixtures

Bathroom #_____ Location _____

TOILET
Model _____
Color _____
Trip Lever _____

SINK(S)
Model _____
Color _____
Spread _____
Faucets _____
Spread   ( ) 4"   ( ) 8"

TUB
Model _____
Color _____
( ) Right Hand   ( ) Left Hand
Size _____ x _____
Faucets _____
_____

WHIRLPOOL
Model _____
Color _____
( ) Right Hand   ( ) Left Hand
Cut sheet attached   ( ) Yes   ( ) No
Access panel location _____
Faucets _____
_____

BIDET
Model _____
Color _____
( ) Right Hand   ( ) Left Hand

Faucets _____

_____

SHOWER
    Shower Pan Material _____

    Enclosure: Model _____ Color_____

     Size_____

ABC BUILDERS
123 Main Street
Hometown, MD 21234
(301) 231-1234

Bathroom #_____   Location_____

FLOOR
Underlayment _____
Tile _____
Supplier _____
Grout _____
Accent pattern or deco tiles_____
_____
_____
Laying instructions, diagonal, etc._____
_____

TUB SURROUND
Wallboard material   ( ) WR Drywall   ( ) Durorock
Tile _____
Supplier _____
Grout _____
Accent pattern or deco tiles _____
_____
_____
Bullnose tile _____
Accessories: Soap dish _____
                    Towel bar _____
Height: To ceiling _____
                Above showerhead _____

SHOWER SURROUND
Wallboard material   ( ) WR Drywall   ( ) Durorock
Tile _____
Supplier _____
Grout _____
Accent pattern or deco tiles_____
_____
_____
Bullnose tile _____
Accessories: Soap dish _____

                    Towel bar _____
        Height: To ceiling _____
                    Above showerhead _____

SHOWER FLOOR
    Material for pan   ( ) mud   ( ) other
    Tile _____
    Supplier _____
    Grout _____

WHIRLPOOL PLATFORM
    Tile _____
    Supplier _____
    Grout _____
    Description or layout including steps
    and backsplash_____

    _____
    _____
    _____
    _____

WAINSCOAT
    Tile _____
    Supplier _____
    Grout _____
    Description or layout including height, bullnose, etc.

    _____
    _____
    _____
    _____

CUSTOMER SERVICE PACKAGE

Pre-Construction Conference Agenda

The Pre-construction conference is the correct place for the builder to educate the buyer on what to expect during the construction process, and to begin to create realistic expectations. This way, when some glitches go wrong (as they inevitably will), it will not result in bad feelings. In order to make the Pre-construction conference as effective as possible, it helps to have an agenda of items the builder will cover during the conference, and room to indicate follow-on tasks that result from your discussion.

The Pre-Construction Conference
Objectives:
- Ensure that home buyers have high levels of trust and confidence in their builder.
- The builder and the buyers are working "off the same page." The buyers' expectations of the construction process, quality, draw procedure, change orders are realistic. Understanding of soil report, foundation type, house plans and specs, anticipated schedule, and other relevant topics are reviewed. A comfortable pattern is established for future communications.
- Remaining questions are identified and noted. A timetable and the responsibility of each party in resolving these items is outlined to avoid misunderstandings. The "put it in writing" habit is established.

Procedures:
Schedule a minimum 2-hour meeting with buyers. Explain that the purpose is to review all plans and information prior to starting construction, and to identify remaining questions. Ask buyers to note any such issues so they do not forget to mention them. Remind the buyers that a site visit will follow the discussion in the office; dress accordingly.
- Assemble all materials necessary for meeting, including appropriate copies for buyers. See Agenda for complete list.
- Note any issues or questions that remain to be discussed or finalized from your point of view.
- Lead the meeting. As applicable, review and discuss each item on the Pre-Construction Meeting form, including issues you noted and items clients bring up.
- Follow-up: For each item, designate who will perform this follow-up and the expected time frame for an update on the topic.

Warranty Forms

Most new homes come with a warranty that covers specified items within a given time frame. While warranty service can be a pain in the neck (especially with unrealistic clients), it provides the consumer the peace of mind that can ease the total building process.

In addition, warranty work, if performed cheerfully and promptly, can create a great deal of positive word of mouth advertising. Studies have shown that the most loyal customer is not the one who never had a problem, but the one who had a problem, but had it promptly fixed. Other studies show that customer satisfaction is directly proportional to the amount of time it takes for the customer to get satisfaction. The longer it takes, the less satisfied they are.

If you have a good warranty and after-sales service program, it can often pay for itself by charging for non-warranty items. If the charges are reasonable, and customer expectations are that they are personally responsible for these routine maintenance items, it should not be an unpleasant experience. If you find it is more trouble than it is worth, find a reliable contractor who will work directly with the customer.

Solution = Effective Warranty Service
Objectives
- The Homeowners feel warranty service is fair, prompt, and effective.
- Warranty requests are resolved within stated time frames, using appropriate subcontractors to control expense.
- The house file contains adequate documentation on work completed in each home.

Procedures:
- The Builder initiates non-emergency warranty service at two points during the warranty year:
  - Sixty days after closing, initiated by the "Welcome" letter sent two weeks after closing with a "one time" repair sheet
  - 11 months after closing.
- Processes non-emergency warranty items per the current Warranty Procedures
- Responds to emergencies reported by phone by the Homeowner
- Responds to miscellaneous service requests, including
  1. warranty claims submitted between the sixty day and eleven month checkpoints
  2. requests for information only
  3. requests for information or service from subsequent owners or real estate agents
- Tracks the number, nature and completion of warranty items: target completion of 90% of routine service requests within 10 business days
- Analyzes warranty items for recurring patterns, design flaws, or other causes that could be eliminated.

ABC BUILDERS
123 Main Street
Hometown, MD 21234
(301) 231-1234

This letter is sent out immediately upon completion of closing. It serves as a cover letter to the service request form and gives basic instructions on the procedures to follow to request service.

Sample Welcome to Warranty Letter

Dear (homeowner):

On behalf of (ABC Builders) I'd like to welcome you to the community once again. We all hope that you are settled and enjoying your new home. While we feel that we delivered an excellent home to you we are realistic enough to recognize that mistakes do happen. Our limited warranty spells out the services we provide in this regard.

If you have any warranty items that need attention at this time, please complete and return the enclosed service request to me on or about the end of the first month in your new home. You may also note any items from your pre-closing inspection that have not been completed just yet.
Upon receiving a service request, I will either contact you for an inspection appointment if one is necessary, or simply issue the appropriate service orders with a copy sent to you as well as the contractor assigned to do the work.

Please feel free to call me if you have any questions.

Sincerely yours,

Warranty Service Rep.

Enclosure: Warranty Service Request

123 Main Street
Hometown, MD 21234
(301) 231-1234

This is sent out two months in advance of the expiration of the one-year warranty items. Enclosed with this letter is a Service Request Form. By giving the homeowner advance warning of the expiration dates, you have indicated to customers that you are serious about your warranty commitments. In addition, where liability for warranty work is carried by suppliers and subcontractors, you avoid a situation where you either have to deal with a disgruntled customer ("You didn't remind me of the expiration") or having to assume liability yourself.

Sample Year End Letter

(Logo)

Inside Address Date

Dear (Homeowner):

It has been over ten months since you closed on your new (ABC Builders) home. We hope you have found your home and the surrounding community to be a pleasant and comfortable place to live.
As you are aware, the Materials and Workmanship portion of your (ABC Builders) Limited Warranty will expire on (date).

If there are any items in your home that require warranty attention, please fill out the enclosed Service Request and return it to our office by (date). Upon receipt of your report we will contact you for an inspection of the items, schedule needed repairs, and answer any questions that you may have.

Sincerely yours,

Warranty Service Rep.

Enclosure: Warranty Service Request

ABC BUILDERS
123 Main Street
Hometown, MD 21234
(301) 231-1234

This is an internal form. All phone messages related to warranty should be recorded here. Where appropriate, a copy of the phone log can be attached to a Warranty Service Request Form that is given to subcontractors or field personnel.

Phone Log

Date/Time _____

Name _____

Address _____ Community _____

Phone/Home _____ Lot _____

Phone/Work _____ Plan _____

Phone/Work _____ Closing Date _____

Date/Message or Item _____ Action/response _____

_____

_____

_____

_____

Follow- up Notes _____

_____

_____

_____

By _____

ABC BUILDERS   **Warranty Service Request Form**
123 Main Street
Hometown, MD 21234
(301) 231-1234

All requests for Warranty Service should be in writing. In an emergency situation, the warranty office can fill out the request form and have the homeowner sign it when the service providers come to the home. Warranty Service Request Forms are provided to the homeowner with the Welcome to Warranty package, and new forms are provided as the forms are returned.

Warranty Service Request

(logo)   ___ 60 day list
Return Address   ___ 11 month list
                                                ___ Emergency follow-up
                                                ___ Other

With the exception of specified emergencies, all requests for service must be in writing. Please use this form to notify us of warranty items. Mail to the address shown above. We will contact you to set an inspection appointment. Service appointments are available from 8 am to 4 pm, Monday through Friday. Thank you for your cooperation.

Name _____

Address _____   Community _____

Phone/Home _____   Lot _____

Phone/Work _____   Plan _____

Phone/Work _____   Closing Date _____

SERVICE REQUESTED _____
_____

SERVICE ACTION _____
_____

COMMENTS _____
_____

Homeowner _____

Homeowner _____

ABC BUILDERS   **One Time Repair Request**
123 Main Street
Hometown, MD 21234
(301) 231-1234

This is a variant of the Warranty Service Request Form. Many items that are not covered under warranty, such as drywall or tile grout repairs, can be a cause of customer · dissatisfaction. The purpose of these one-time repairs is to show the homeowner how to do the maintenance in the future, and to educate the homebuyer that these items are NOT an on going maintenance responsibility of the homebuilder.

"One Time" Repairs

Just as some car dealers provide the first oil change when you purchase a new car, we provide several "first time" repairs for your home. Your Homeowner Manual lists these under individual headings (for instance drywall and grout) in the "Caring for Your Home" section. This service is provided as a courtesy and to give you an opportunity to observe methods and materials needed for ongoing maintenance of your home. Only ONE "One Time" Repair request per home, please! We suggest sending this in near the end of your warranty year to maximize the benefits you receive. Simply complete and mail this form to our office with your year-end warranty list. Thank you!

Name _____

Address _____   Community _____

Phone/Home _____   Lot _____

Phone/Work _____   Plan _____

Phone/Work _____   Closing Date _____

SERVICE REQUESTED _____

_____

SERVICE ACTION _____

_____

_____

COMMENTS _____

_____

Homeowner _____

Warranty Work Order

This is the form sent to the subcontractor authorizing the work to proceed and payment to be made upon completion. It is to be completed by your Customer Service staff after receiving a Warranty Service Request Form. It establishes responsibility (the subcontractor is responsible for making the service appointment) and also establishes a time line for completion of work (within 10 working days of receipt of the form). A copy of this Warranty Work Order is returned to the Customer Service office upon completion of the work. Customer Service should then follow up with the homeowner to ensure their satisfaction before paying the work order.

ABC BUILDERS
123 Main Street
Hometown, MD 21234
(301) 231-1234

Warranty Work Order

Date _____   Community _____

Work Order # _____   Lot # _____

Payment Approved _____   Model _____

Subcontractor _____   Homeowner _____

Address _____   Address _____

Phone _____   Phone (H) _____

Phone (W-Ms.) _____

Phone (W-Mr.) _____

Work Requested: _____
_____
_____
_____
_____
_____
_____
_____
_____
_____

Requested by: _____
Completed by: _____
Date: _____

The homeowner has received a copy of this service order and will expect this work to be completed within 10 business days. It is your responsibility to set up a service appointment, although the homeowner may call you to expedite this. Upon completion of the work, sign and return this form for the warranty file. Your attention and cooperation are appreciated!

Comments on work performed: _____
_____
_____
_____
_____

ABC BUILDERS
123 Main Street
Hometown, MD 21234
(301) 231-1234

Work Order Log

All customer service work orders should be logged in one place. This is to ensure that there are no missing work orders. It also provides a good place to track uncompleted work, without requiring the staff to go through a stack of work orders.

Contractor _____

Phone _____   Contact _____

Date Issued	Work Order #	Homeowner	Comment	Days to Complete

Copy to contractor on _____, 19 _____

OFFICE MANAGEMENT PACKAGE

ABC BUILDERS   **Fax Cover Sheet**
123 Main St.
Hometown, MD 21234
(301) 231-1234

Something as mundane as a well designed fax cover sheet can present an improved image for your company, avoid confusion, and make it easier for your clients and business associates to communicate with you. A good fax cover sheet should always:
1. Be graphically representative of your company. Just like signage, stationery, etc., all materials which represent your business should be consistent and should create a strong graphic image.
2. Give your fax number for a reply.
3. Give your telephone number in case of a problem in transmission.
4. Give your address in case the recipient wants to mail something to you.
5. Give the name of the person sending the fax.
6. Give the name of the person for whom the fax is intended.

ABC BUILDERS
123 Main St.
Hometown, MD 21234
(301) 231-1234

Fax

To:		From:	
Fax:		Pages:	
Phone:		Date:	
Re:		CC:	

☐ Urgent   ☐ For Review   ☐ Please Comment   ☐ Please Reply   ☐ Please Recycle

Comments: _____

_____

_____

_____

_____

_____

_____

_____

_____

_____

_____

_____

_____

ABC BUILDERS
123 Main St.
Hometown, MD 21234
(301) 231-1234

A transmittal sheet serves the same purpose as a fax cover sheet, but is for documents sent through the mail.

Transmittal Sheet

To:		From:	
Fax:		Pages:	
Phone:		Date:	
Re:		CC:	

☐ Urgent   ☐ For Review   ☐ Please Comment   ☐ Please Reply   ☐ Please Recycle

Comments: _____
_____
_____
_____
_____
_____
_____
_____
_____
_____
_____
_____

ABC BUILDERS
123 Main St.
Hometown, MD 21234
(301) 231-1234

These are the minimum supplies you should have in a your sales office. It's useful for setting up a new office, or for ordering supplies for an existing one.

Date _____

Your sales office should be equipped with the following materials:

Marketing Materials
- [ ] 4 page marketing brochure
- [ ] 2 page marketing flyer
- [ ] "10 Reasons" flyer
- [ ] Direct Mail Postcard
- [ ] Modular information/models/developments
- [ ] 3 ring binder with samples of current ads
- [ ] Photo album of past projects

Customer Contact Forms
- [ ] Realtor Registration Form
- [ ] Request for Information Card
- [ ] Prospect Card
- [ ] Referral Card

Client Forms
- [ ] Client Selection Form
- [ ] Standard Charges/Prices
- [ ] Kitchen Design Worksheets
- [ ] Bath Design Worksheets

Contract Forms
- [ ] Contract Acceptance Letter
- [ ] Custom/Production Home Buyer Contract
- [ ] Standard Allowance Schedule
- [ ] Design Contract
- [ ] Professional Services Agreement
- [ ] Specifications Form

Office Supplies
- [ ] Ballpoint pens
- [ ] Felt-tip pens
- [ ] Lined writing tablets
- [ ] Toner cartridges/copier
- [ ] Toner/inkjet supplies/printer
- [ ] Copier paper
- [ ] Computer diskettes
- [ ] Paper clips
- [ ] Scotch tape
- [ ] 3-ring binders
- [ ] File folders
- [ ] Fax paper

Office fixtures
- [ ] Literature display (take one)
- [ ] Holder for forms/materials
- [ ] Desk
- [ ] File cabinet
- [ ] Conference table with 4 comfortable chairs
- [ ] Comfortable desk chair
- [ ] Personal computer
- [ ] Copier
- [ ] Computer printer (color inkjet)
- [ ] 17" color monitor
- [ ] Fax/telephone/answering machine
- [ ] Builder story display
- [ ] Bookcase (reference materials)
- [ ] 3-hole punch

123 Main St.
Hometown, MD 21234
(301) 231-1234

These are the important things to look for in reviewing existing marketing materials, or in preparing new ones. They are divided into 3 sections: The Message, The Media, and the Means. Of these, the message is the most important area, and accounts for most of the impact of your marketing effort. Of course, if you have the right message, but nobody sees it, it won't do much good. But if you have the wrong message, you can spend a great deal of time and energy getting the message out, and have it fall upon deaf ears. The means is the resources (time and money) that you devote to marketing. As a general rule of thumb, we advocate that small builders spend between 1½% and 2% of gross sales for marketing. This does not include sales commissions, but does include the creation and distribution of your marketing materials.

Part of the message is your Value Proposition. There are five distinct value propositions a builder can use. However, he must consistently convey the same proposition in order to avoid confusing buyers. Pick one and then consistently apply it in all materials. These 5 propositions are:

_____ Gives more for the same price
_____ Gives the same at a lower price
_____ Gives more at a lower price
_____ Gives a little more at a much lower price
_____ Gives a lot more for a slightly higher price

In evaluating materials, it's more important that they convey the right message and are benefit-oriented, rather than being glitzy or expensive. However, they should be professional looking, and convey an image of quality. Second and third generation copies send a subliminal message of lack of concern. If you need to, pay an outside artist or writer to create your materials, and then reuse those images and words in all your marketing efforts.

Marketing Materials Checklist

The Message

1. ☐ Do all marketing materials focus on benefits (what the product does *for* the buyer), rather than on features (what the product is or does)?
2. ☐ Do all marketing materials focus on 3 or 4 major benefits for buyers?
3. ☐ Do all materials have a consistent image? Are they clean and easy to read?
4. ☐ Do we have a strong company image, including logo, and theme? Is it used consistently on all marketing materials?

5. ☐ Do we use all the attributes of the product – emotional, procedural, and functional, as well as physical and financial?
6. ☐ Do we have a consistent value proposition that defines how we relate to our competition in terms of quality and price?
7. ☐ Do we use interior photos that allow the buyer to imagine what it would be like to live in our product?
8. ☐ Do we use captions beneath photos to reinforce the message?
9. ☐ Do we use testimonials to create credibility?
10. ☐ Do all materials contain a call to action, including a phone number prominently displayed?

The Media
Onsite
Do we have the following materials available:
1. ☐ 4-page (or more) company positioning brochure
2. ☐ "10 reasons to buy" flyer
3. ☐ Modular materials on individual model homes and communities
4. ☐ Information holder
5. ☐ Other _____

Direct Mail
1. ☐ 1 page flyer
2. ☐ 2-page brochure
3. ☐ Postcard mailer
4. ☐ Special events mailers
5. ☐ Other _____

Space Advertising
1. ☐ 1-page ad (newspaper)
2. ☐ ½ page ad (newspaper)
3. ☐ ¼ page ad (newspaper)
4. ☐ 1 page ad (magazine)
5. ☐ ½ page ad (magazine)
6. ☐ ¼ page ad (magazine)
7. ☐ Other _____

Other Media
1. ☐ Internet site
2. ☐ Open House
3. ☐ Parade of Homes
4. ☐ Display booth
5. ☐ "Take one" Boxes
6. ☐ On-site Signage
7. ☐ Other _____

Means

Percent of gross sales allocated to marketing _____

Annual Man-hours/month devoted to marketing _____

Person with primary responsibility for marketing results _____

PROFIT MANAGEMENT PACKAGE

Month End Reporting

Every builder should do a thorough financial review every month. These are the financial reports we suggest looking at.

1. **Income Statement Highlights**
   - Include Month to Date and Year to Date Totals
   - Include Budget and Actual Amounts
   - Report on the following main categories:
     - Sales and Gross Profit for each main category of work performed
     - Overhead Expense by Major Category

2. Detail Operating Expenses
   - Include Month to Date and Year to Date Totals
   - Include Budget and Actual Amounts

3. Summarized Job Analysis
   - Revenue and Costs should tie to the Income Statement
   - Separate completed jobs from work in progress
   - Include original contract, change orders, costs to date, costs to complete, billings to date, revenue earned, over (under) billings and gross profit remaining

4. Balance Sheet Highlights
   - Include columns for current month, prior month and change
   - Highlight changes in working capital, fixed assets and long term debt

5. Key Ratios
   - Monitor monthly changes and identify trends
     - Working capital ratio (current assets/current liabilities)
     - Debt to equity ratio (total liabilities/total equity)
     - Months overhead covered (operating expenses compared to gross profit remaining on jobs in progress)

Income Statement

The Income Statement is the most important financial report the builder must examine. It is a summary of the revenue, expenses and net income (earnings) or loss. Net income or loss becomes part of the Balance Sheet by increasing (net income) or decreasing (net loss) the owner's capital.

Proper analysis of the Income Statement will assist a builder in evaluating the overall profitability of his business. In the upper part of the income statement the overall amount of revenue and costs relating to jobs is shown. The builder should keep a good eye on gross profit margins when planning to price future jobs.

The lower portion of the Income Statement reflects the expenses that were incurred in the operation of the business. These include financing and overhead expenses. In order to get the most out of the Income Statement, the lower portion may be expanded to include a more detailed analysis of operating expenses. In addition to monitoring the percentage relationships that the various expense categories have to revenue, the Income Statement can be used to compare budgeted revenue and expenditures against actual sales and expenses. A sample Income Statement is shown.

ABC BUILDERS
123 Main Street
Hometown, MD 21234
(301) 231-1234

Income Statement

Sample

ABC BUILDERS - INCOME STATEMENT
JUNE 1 - JUNE 30, 1996

MONTHLY

	BUDGET	ACTUAL	VARIANCE	SALES	BUDGET % SALES
Revenue	$ 200,000	225,000	25,000	100%	
Costs of Construction	168,000	192,750	(24,750)	84.0%	85.7%
Gross Profit	$32,000	32,250	250	16.0%	14.3%
Overhead					
Sales & Marketing	1.400	1,102	298	0.7%	0.5%
Salaries & Fringe Benefits	10,180	10,410	(230)	5.1%	4.6%
Office Expenses	2,657	2,443	214	1.3%	1.1%
Vehicle, Travel & Indirect	815	1,188	(373)	0.4%	0.5%
Other Overhead	1,324	2,887	(1,563)	0.7%	1.3%
Total Overhead	16,376	18,030	(1,654)	8.2%	8.0%
Income (Loss) From Operations	15,624	14,220	1,404	7.8%	6.3%
SALES AND MARKETING					
Advertising	200	275	(75)	0.1%	0.1%
Marketing & Public Relations	1,100	599	501	0.6%	0.3%
Customer Entertainment	100	228	(128)	0.1%	0.1%
Total Sales and Marketing	1.400	1,102	298	0.7%	0.5%
SALARIES AND FRINGE BENEFITS					
Owners' Salaries	4,000	4,000	-	2.0%	1.8%
Administrative Salaries	4,600	4,750	(150)	2.3%	2.1%
Payroll Taxes	688	732	(44)	0.3%	0.3%
Workers' Compensation Insurance	117	139	(22)	0.1%	0.1%
Health and Accident Insurance	775	789	(14)	0.4%	0.4%
Total Salaries & Benefits	10,180	10,410	(230)	5.1%	4.6%
OFFICE EXPENSES					
Office Rent	1,000	1,000	-	0.5%	0.4%
Equipment Rent	75	75	-	0.0%	0.0%
Office Repairs and Maintenance	45	0	45	0.0%	0.0%
Equipment Repairs & Maintenance	25	150	(125)	0.0%	0.1%
Utilities	260	342	(82)	0.1%	0.2%
Communications	545	635	(90)	0.3%	0.3%

Office Supplies	425	168	239	0.2%	0.1%
Postage	32	55	(23)	0.0%	0.0%
Computer Expenses	250	0	250	0.1%	0.0%
Total Office Expenses	657	2,443	214	1.3%	1.1%

VEHICLE, TRAVEL AND INDIRECT EXPENSES

Vehicle Lease	315	315	-	0.2%	0.1%
Vehicle Repairs and Maintenance	0	329	(329)	0.0%	0.1%
Gas and Oil	200	189	11	0.1%	0.1%
Taxes, License and Insurance	200	205	(5)	0.1%	0.1%
Travel	0	0	0	0.0%	0.0%
Indirect Job Expenses	100	150	(50)	0.1%	0.1%
Total Vehicle, Travel & Entertainment	815	1,188	(373)	0.4%	0.5%

OTHER OVERHEAD EXPENSES

Miscellaneous Taxes	-	-	-	0.0%	0.0%
Package Insurance	0	1,600	(1,600)	0.0%	0.7%
Accounting	175	175	0	0.1%	0.1%
Legal	200	0	200	0.1%	0.0%
Depreciation	750	750	0	0.4%	0.4%
Contributions	25	0	25	0.0%	0.0%
Dues and Subscriptions	40	36	4	0.0%	0.0%
Education and Training	25	50	(25)	0.0%	0.0%
Interest Expense	109	276	(167)	0.1%	0.1%
Total Other Overhead Expenses	1,324	2,887	(1,563)	0.7%	1.3%

Total General and Administrative Expenses	14,976	16,928	(1,654	7.5%	7.5%

Balance Sheet

The Balance Sheet is a statement showing the assets, liabilities and equity of a business entity at a specific date. Assets are those items that are owned by the business, whereas liabilities include items that the business owes. Equity is the owners' interest in the enterprise. The Balance Sheet tells the business owner things like how much cash he has in the bank, how much money is owed to him and what he owes out, how much equipment is on hand and what the business is worth at a given point in time.

Proper analysis of a Balance Sheet will help a builder understand and evaluate his financial strength, liquidity and leverage. It will also provide information that will help in the formulation and monitoring of company goals and plans. Comparative analyses of more than one Balance Sheet can assist you in understanding your company's structural changes and liquidity patterns. A sample Balance Sheet is provided.

ABC BUILDERS
123 Main Street
Hometown, MD 21234
(301) 231-1234

Sample

BALANCE SHEET

ASSETS

CURRENT ASSETS
Cash	$ 2,750
Accounts receivable	50,000
Construction in progress (Spec Home)	106.900
Prepaid insurance	4,230
Other current assets	3,426
Cost and earnings in excess of billings	25,847
Total Current Assets	$ 193,153

PROPERTY & EQUIPMENT
Construction Equipment	$ 21,225
Office Equipment	28,700
Vehicles	34,500
Computer Equipment	10,230
	94,655
Less: Accumulated Depreciation	55,525
	39,130
Total Assets	232,282

LIABILITIES & STOCKHOLDERS' EQUITY
Current Liabilities	$ 10,000
Note payable -- line of credit	3,125
Current portion of long term debt	126,570
Accounts payable	4,903
Payroll taxes withheld and accrued	64,777
Billings in excess of costs and earnings	$ 209,465
Total Current Liabilities	
Long term debt	3,240

Stockholders' Equity

Common stock: $1 stated value. 1,000 shares	1,000
Retained Earnings	
Balance, January 1, 1996	15,865
January - June Net Income	2,713
Balance, June 30, 1996	<u>18,578</u>
	<u>19,578</u>
Total Liabilities and Stockholders' Equity	$ 232,283

Ratio Analysis

By monitoring certain key ratios you can get a better understanding of where you've been and where you're going. Two ratios which are important for a builder to monitor are the current ratio and debt to equity ratio.

Current Ratio

The current ratio is the most popular measure of a company's solvency. It compares the amount of current assets with which payments can be made to the amount of current liabilities requiring payment. Analysis of this ratio will indicate shortages in working capital or a reluctance to utilize available resources. The higher the ratio, the more capable the company is in meeting its current obligations.

$$\text{Current Ratio} = \frac{\text{Current Assets}}{\text{Current Liabilities}}$$

Debt to Equity Ratio

The debt to equity ratio is a key measure of the company's leverage factor. Banks strongly rely on this ratio in their evaluation of a builder's creditworthiness. The higher the ratio, the more risk is being assumed by creditors. A lower ratio usually indicates that the company has more borrowing capacity and greater long-term financial stability.

$$\text{Debt-To-Equity Ratio} = \frac{\text{Total Liabilities}}{\text{Total Owners' Equity}}$$

Job Cost Analysis

While the Income Statement assists a builder in evaluating the overall profitability of his jobs, it does not provide the builder with an analysis of the profitability on a job-by-job basis. The primary purpose of preparing and evaluating Job Cost reports is to provide reports on a timely basis which assist you in controlling and managing construction costs.

At a minimum, Job Cost reports can be used to compare estimated to actual expenditures for various areas of work performed on each job. More sophisticated Job Cost reports can be used to also take into account projected costs to complete as well as analyze the status of labor and subcontract costs.

Regular preparation and analysis of Job Cost reports can be a valuable tool in helping to identify and correct problems which will lead to cost overruns before it is too late.

Cash Flow Forecasting

Did you know that the construction industry accounts for the second largest number of bankruptcies next to mom and pop restaurants? Did you also know that the number one reason for failure among builders is not lack of profitability but poor cash flow? One way of improving your odds of survival and increasing your profitability is by cash flow forecasting. Cash flow forecasting is the planning of cash availability and requirements for a specified period of time.

Cash flow forecasting can be beneficial in a number of ways. It can provide you with a means to evaluate growth and changes in the business. In addition, it provides a method of identifying the real cost of money as it relates to time.

For your banker and yourself, cash flow forecasting reflects a sophisticated level of organization and planning which can lead to increased credit and reduced interest rates. Cash flow forecasts should be a part of the overall budgeting and planning functions. Forecasts can be prepared both on a project-by-project basis and an overall company basis for both long (annual) and short (monthly and weekly) periods of time.

Construction Cash Flow Projections

As part of the planning process for each construction project, the builder should identify the amount and timing of cash needed throughout the job. The cash flow projection should be a by-product of the scheduling and estimating functions. By knowing estimated project costs, when they are anticipated to be incurred, the payment terms for materials and subcontractors and the billing and collection terms that you have with the customer, a cash flow forecast for the project can be easily prepared.

Company-Related Cash Flow Projection

After completing cash flow projections for each project, the net cash available or required from each project can be summarized and used to prepare an overall company-related cash projection. This projection consolidates construction cash requirements with company overhead expenses.

By preparing and monitoring cash flow projections, the builder truly is looking into a crystal ball. What the crystal ball is saying can be essential to the profitability and ultimate survival of your company.

Sample Builder
Monthly Cash Flow Projection

Description	January	February	March	April	May	June	Totals
Beginning Cash Balance	$15,000	$64,903	$34,791	$(911)	$12,818	$(7,137)	
Cash Sources							
Draws	88,000	25,000	117,000	104,100	145,500	80,000	559,600
Sales	189,900	-	-	210,000	-	-	399,900
Other	-	2,250	-	-	-	-	2,250
Total Cash Sources	277,900	27,250	117,000	314,100	145,500	80,000	961,750
Cash Outflows							
Construction Costs	50,000	43,000	130,000	97,500	150,000	88,000	558,500
Overhead	12,497	13,862	14,702	14,871	14,955	13,406	84,293
Other	165,500	500	8,000	188,000	500	500	363,000
Total Cash Outflows	227,997	57,362	152,702	300,371	165,455	101,906	1,005,793
Net Cash Flow	49,903	(30,112)	(35,702)	13,729	(19,955)	(21,906)	(44,043)
Ending Cash Balance	$64,903	$34,791	$(911)	$12,818	$(7,137)	$(29,043)	

Cash Sources

Description	January	February	March	April	May	June	Totals
Draws							
Lot 123-Loan	$20,000	$25,000	$20,000	$ -	$ -	$ -	$65,000
Lot 16-Loan	-	-	36,000	24,000	50,000	10,000	120,000

							Total
Model Home-Loan	18,000	-	-	-	-	-	18,000
Mr. & Mrs. Smith	-	-	36,000	30,100	20,500	50,000	136,600
Mr. & Mrs. Silver	50,000	-	25,000	50,000	75,000	20,000	220,000
Total Draws	$88,000	$25,000	$117,000	$104,100	$145,500	$80,000	$559,600

Sales

							Total
Lot 125-Spec	$189,900	$ -	$ -	$ -	$ -	$ -	$189,900
Lot 123	-	-	-	210,000	-	-	210,000
Total Sales	$189,900	$ -	$ -	$210,000	$ -	$ -	$399,900

Other Cash Sources

							Total
Other Income	$ -	$ -	$ -	$ -	$ -	$ -	$ -
Insurance Refund	-	2,250	-	-	-	-	2,250
Total Other Cash	$ -	$2,250	$ -	$ -	$ -	$ -	$2,250

							Total
Total Cash Sources	$277,900	$27,250	$117,000	$314,100	$145,500	$80,000	$961,750

Cash Outflows

Construction Costs

							Total
Lot 123	$25,000	$30,000	$25,000	$ -	$ -	$ -	$80,000
Lot 16	-	5,000	30,000	30,000	55,000	8,000	128,000
Model Home	20,000	3,000	-	-	-	-	23,000
Mr. & Mrs. Smith	-	-	35,000	27,500	25,000	60,000	147,500
Mr. & Mrs. Silver	5,000	5,000	40,000	40,000	70,000	20,000	180,000

							Total
Total Construction Costs	$50,000	$43,000	$130,000	$97,500	$150,000	$88,000	$350,500
Operating Expenses							
Salaries and Fringes	$8,620	$8,560	$10,200	$8,080	$9,778	$7,984	$53,222
Sales and Marketing	325	1,125	325	825	1,250	675	4,525
Vehicle & Indirect	790	1,140	790	1,065	1,215	790	5,790
Office Expense	2,182	2,432	2,307	3,432	2,182	3,407	15,942
Other Expenses	580	605	1,080	1,469	530	550	4,814
Total Operating Expenses	$12,497	$13,862	$14,702	$14,871	$14,955	$13,406	$84,293
Other Cash Outflows							
Repay Construct Loan							
Lot 125-Spec	$150,000	$ -	$ -	$ -	$ -	$ -	$150,000
Lot 123	-	-	-	170,000	-	-	170,000
Closing Costs							
Lot 125-Spec	15,000			-			15,000
Lot 123	-	-	-	17,500	-	-	17,500
Vehicle Note Payments	500	500	500	500	500	500	3,000
Corporate Tax Estimate	-	-	7,500	-	-	-	7,500
Total Other Cash Outflows	$165,500	$500	$8,000	$188,000	$500	$500	$363,000
Total Cash Outflows	$227,997	$57,362	$152,702	$300,371	$165,455	$101,906	$797,793

Yearly Marketing Budget

The correct answer to how much you should spend is whatever it takes to get the results you want. If you don't want any results you don't have to spend any money. All other things being equal, given a good message and good media placement, the more you spend on marketing, the more sales will result. We generally recommend spending 1½-2% of your gross sales for marketing, not including sales commissions. If you want to increase sales, spend 2% of the sales volume you would *like* to generate.

There are two aspects of the marketing budget – discretionary costs and non-discretionary costs. Non-discretionary costs are the minimum costs of setting up a program. These include creating the message, graphic design, stationery, business cards, and signage. We generally recommend that you budget $3,000 to $4,000 for this work. If you already have a good logo and graphic identity, then you can save about $1,000 of this. DO NOT SCRIMP on the non-discretionary part of the budget, since the effectiveness of the entire campaign is dependent on the quality of the message and materials. Discretionary costs are a function of the level of sales you intend to generate. This includes additional copies of the brochures and flyers for direct mail, postcards, and space advertising in newspapers and newsletters.

There are several ways of reducing the cost of producing good marketing materials. One way is to list your major subcontractors as part of the building team, and then charge each contractor a small fee ($100-200) for the exposure. If you have 15 subcontractors and each pays $200, that's $3,000, which could cover the printing costs of a 4 page 4-color 5,000-run brochure. The other method is to use co-op funds obtainable through your local suppliers. If you use Andersen windows exclusively, for example, they may be willing to subsidize part of the production costs for a small paragraph (with logo) promoting their products. Typically, co-op fees can pay up to half of production costs. Or you can combine both methods and essentially get marketing materials for free.

ABC BUILDERS   **Yearly Marketing Budget**
123 Main Street
Hometown, MD 21234
(301) 231-1234

Message and Graphic Identity $ 3,000
Includes design of logo, stationary, business cards,
  signage and creation of marketing message.

Brochure (1,000 copies)* $ 3,250
Flyer (2,500 copies) – 4 color* $ 3,000
Post card (4,000 copies) $ 2,000
Mailing $ 1,200
Ads – local paper 12 @ $400* $ 4,800
Ads – targeted media 10 @ 250 $ 2,500
Signs – 6 @ $125 $ 750
Business cards/stationary $ 350
2 open houses (1 custom/1model/spec) $ 2,500
Gifts/promotions $ 1,250

TOTAL $24,350

* Less $1,000 suppliers, $2,400 coop   <3,400>

                                                                    $20,950

TOTAL BUDGET $21,300

123 Main Street
Hometown, MD 21234
(301) 231-1234

Management by Exception is the concept that you put your effort in those areas that aren't operating as expected (exceptions). Your system should be designed to automatically flag all items which exceed their budgets by more than a stipulated percent, and all construction tasks which are behind schedule. This allows you to spend your time solving problems that need to be solved, and that are draining your company of profits.

Sample Builder
Weekly Exception
Report

Purchase Order
Variances

Job	Cost Code	Variance Amount	Variance Reason

Schedule Variances

Job	Task Code	Days Affected	Variance Reason

ABC BUILDERS
123 Main Street
Hometown, MD 21234
(301) 231-1234

If your accounting system is set up to report on a timely basis, you should review expenses and income on a weekly basis. A weekly status report should give you your current balances, the status last week, and the change during the week. This way you can quickly pinpoint problem areas, and get more detailed reports in that area.

Sample Builder
Weekly Status Report

	Current Balance	Last Week	Change
Cash	7,422	$2,823	4,599
Accounts Payable			
Active Jobs			-
Closed Jobs			-
Overhead			-
Other Payables			-
Total Payables	-	-	-

Work in Progress Status

	Current Balance	Last Week	Change
Direct Construction Costs			-
Less:			
Construction Loans			-
Customer Deposits			-
Active Job Payables			-
Net Work in Progress	-	-	-

Cash Activity

Beginning Balance		$7,422
Cash Receipts		
Loan Draws	$5,000	
Customer Payment	10,000	
Sales	-	
Miscellaneous Deposits	234	
Total Cash Receipts		15,234
Cash Disbursements		
Loan Payments	893	
Payment of Job Expenses	15,240	
Payroll and Taxes	2,500	
Opertating Expenses	1,200	
Total Cash Disbursements		19,833
Ending Balance		$2,823

ABC BUILDERS
123 Main Street
Hometown, MD 21234
(301) 231-1234

In order to improve your operations, you have to understand what recurrent problems you have, and devise strategies to fix them. These are problems common to many builders, and by putting your finger on them, you can start to do something about them.

Date _____

Below are 30 common problems faced by small builder and contracting firms. They are divided into five functional areas. Review all of these issues, and check all those that you think could use improvement in your company.

FINANCE/CONTROL
- ☐   1.   Inadequate or non-existent business plan.
- ☐   2.   Lack of operating budget and cost controls.
- ☐   3.   Poor capitalization
- ☐   4.   Low (or unknown) profit margin.
- ☐   5.   Inaccurate and inefficient estimating procedures

PRODUCT DESIGN
- ☐   6.   Product lacks pizzazz -- insufficient market appeal.
- ☐   7.   Product is expensive to build. Too complicated.
- ☐   8.   No Design/Build capability - Inability to customize existing designs.
- ☐   9.   Lack of separation and identity in marketplace. "Me too" designs.
- ☐   10.   Failure to introduce new products, features, ideas into home design

ADMINISTRATION
- ☐   11.   Lack of skill or comfort negotiating with subs and clients
- ☐   12.   Personnel problems: high turnover, poor productivity.
- ☐   13.   Inefficient administrative procedures: scheduling, payroll, paperwork
- ☐   14.   Poor purchasing controls
- ☐   15.   Weak company programs: job site safety, health, benefits

CONSTRUCTION
- ☐   16.   Lack of skilled craftsmen, reliable subcontractors
- ☐   17.   Lack of written specifications and contracts with subcontractors
- ☐   18.   No formal scheduling system
- ☐   19.   Quality control problems
- ☐   20.   Lack of timely response to material and manpower needs

SALES AND MARKETING

☐   21.   Non-existent or ineffective marketing strategy
☐   22.   Inadequate marketing materials (brochures, ads, logo, etc.)
☐   23.   Low referral rate from previous clients, designers, and other professionals
☐   24.   Ineffective or non-existent referral program with local realtors
☐   25.   Low closing rate with prospective clients

CUSTOMER SERVICE

☐   26.   Inadequate or non-existent customer service program
☐   27.   Houses not completely ready at final walk-though
☐   28.   Client specification decisions not made in a timely manner
☐   29.   Unrealistic quality expectations by clients
☐   30.   High service warranty costs

PROBLEM PRIORITIES

After you have selected those areas that could stand improvement, then select the three issues where you feel your company is the strongest, and three areas where you feel your company is the weakest. Select areas that you feel would have the greatest positive impact on your company.

THREE GREATEST STRENGTHS

1.

_____

2.

_____

3.

_____

THREE BIGGEST PROBLEMS

1.

_____

2.

_____

3.

_____

Now go through the SOS (Situation/Options/Solutions) problem-solving process for each of these three problems. Identify set deadlines for resolving the problems, and assign responsibility and resources. Then repeat for the next set of three problems.

PROBLEM	SOLUTION	DEADLINE
1. _____	_____	_____
2. _____	_____	_____
3. _____	_____	_____

4. _____   _____   _____

5. _____   _____   _____

6. _____   _____   _____

7. _____   _____   _____

ABC BUILDERS   **Year End Planning Checklist**
123 Main Street
Hometown, MD 21234
(301) 231-1234

At the end of each year, builders should take stock of their progress and measure their achievements for the year against their goals. They should then determine new goals for the upcoming year. Goals should be quantifiable and achievable. Instead of stating "We want to increase sales next year," you should state ""We will increase sales by ___ %." You should then outline the steps you will take to ensure that the goal is achieved.

Look at every aspect of the business. This includes financial management, construction management, office and employee management, customer management, and product management. Each of these areas is interactive. A better product can lead to increased sales, and therefore more profits. Better construction management can lead to lower costs and therefore more profits.

Some sample goals are outlined below: Pick and choose the ones that you feel are applicable to your situation.

Financial Management

- ☐ Create a better financial management system: Include weekly exception reports, monthly profit reviews.
- ☐ Move to integrated computer financial management system with construction scheduling and job-cost reports.
- ☐ Institute a purchase order system to control expenditures.
- ☐ Develop cash management system to forecast and control cash levels.
- ☐ Develop long-term sources of financing, including banks and outside investors.
- ☐ Examine tax liability strategies with your CPA. What can we do to minimize tax liability now?

Customer Management

- ☐ Review and improve current marketing materials.
- ☐ Develop new materials where needed. Pay special attention to the message. Is it benefit-driven and focused?
- ☐ Revamp sales techniques to improve closing ratios.
- ☐ Develop list of common objections and strategies for overcoming them.
- ☐ Implement customer service system -- to eliminate sources of customer dissatisfaction and improve responsiveness to problems as they arise.
- ☐ Evaluate current customer service system. Is it meeting its goals for customer satisfaction?
- ☐ Increase marketing budget to increase market awareness.

☐   Implement proactive referral system using past and present buyers.

Construction Management

☐   Evaluate estimating procedures. Systematize estimating to reduce errors and produce more timely estimates.

☐   Implement or improve construction scheduling system. Develop standard system (computer or manual) that allows you to track schedule more accurately, and compensate for time delays that may occur.

☐   Evaluate house plans for construction efficiency. Simplify construction where possible.

Office and Employee Management

☐   Evaluate employee compensation system. Does it generate the results you want. Are there non-financial recognitions built into the system to improve employee morale and productivity?

☐   Evaluate office productivity. Can it be improved through new computer systems or software? Is additional staffing needed? Would additional staff be able to generate additional profits?

☐   Examine office forms and reporting procedures. Simplify where possible, and eliminate all unnecessary reports and forms.

☐   Evaluate fringe benefits, such as health care, retirement plans, and profit sharing. Are you getting the benefits you're paying for? Are other plans available that would give more benefits for the same cost?

☐   Evaluate your insurance plans. Are you adequately covered for liability? Are your vehicle insurance rates reasonable?

Product Management

☐   Evaluate your current house plans. How do they compare to your competition? Do they have adequate curb appeal? Are the floor plans efficient and exciting?

☐   Go through each house, and find ten details that could be added to increase the "wow" factor without substantially increasing costs.

☐   Evaluate the housing market. Are we building the right house for the right customer? What market niches are we building for? What markets should we be in? How is the market changing in the future?

☐   Evaluate the mix of spec and custom homes. Can we build additional spec homes without dramatically increasing our risk?

Strategic Planning Form

At least once a year, every builder should sit down and look at his operations over the past year and ask himself some very basic questions. Who are we? Where are we going? What did we do right? What did we do wrong? How can we get better?

Honest answers to these questions can help a builder devise a more productive strategy for competing effectively in the marketplace, ensuring a smoothly running operation and increasing profits.

In filling out the strategic planning form, try to be as honest as possible. Wishful thinking does not lead to increased profits. If your competitor is better than you are in certain aspects, then learn from him, and use that as a benchmark to improve your own performance.

Also, try to be realistic. Yes, it would be nice to double your sales next year. But, if you don't have the resources to handle the growth and fund the marketing, and don't have the manpower to deal with the increased workload, then you shouldn't even try.

The most important part of the strategic planning form is the last part, goals. All the thinking and planning in the world won't help you unless you can translate that thinking into concrete goals and actions. Write down all the specific things you want to accomplish in the next year in order to increase your competitiveness and profitability. Then list them in order of priority. Set a date to accomplish each task, and then periodically check with yourself to see how you're doing in getting those tasks accomplished. As soon as you finish your first set of goals, give yourself some new goals and keep moving ahead.

ABC BUILDERS
123 Main Street
Hometown, MD 21234
(301) 231-1234

Strategic Planning Form

Date _____

1. Who are we?

❑ Custom Builder ❑ Semi-Custom Builder ❑ Spec Builder ❑ Production Builder
❑ Land Developer ❑ Builder/Developer ❑ Other _____

How long have we been in business? _____
Annual Sales (over past 5 years in reverse chronological order)
Anticipated this year _____ Last year _____
2 years ago _____ 3 years ago _____
4 years ago _____ 5 years ago _____

Annual Profits (over past 5 years)
Anticipated this year _____ Last year _____
2 years ago _____ 3 years ago _____
4 years ago _____ 5 years ago _____
Number of Employees: _____

Positioning:
❑ We build the best (and most expensive) houses in the area.
❑ Our prices and our quality are comparable to our competition
❑ We build comparable houses for less money than our competition
❑ We build much better houses at the same price as our competition
❑ We build the least expensive houses in the area.

Comments: _____
_____
_____
_____
_____

2. Who are our customers? (Choose all that apply)

By Lifecycle: ❑ Singles ❑ First Married ❑ Growing Family ❑ Extended Family
❑ Empty Nesters ❑ First time buyers ❑ First time move-ups
❑ Second move-ups
By Income ❑ Over $60,000 ❑ Over $75,000 ❑ Over $100,000 ❑ Over $150,000
❑ Over $200,000

By Profession: _____

By Geography: (City or Zip codes) _____

By housing style: ☐ Traditional ☐ Contemporary
                  ☐ Traditional Exterior/Contemporary Interior

By Special Housing Needs: ☐ Large Family ☐ Small Family ☐ No Kids
                          ☐ Handicapped Access ☐ 3 – 4 Car Garage ☐ Home Office
                          ☐ Entertaining Space ☐ Home Theatre ☐ Pro. Kitchen

Other _____

_____

Comments: _____

_____

_____

_____

_____

_____

_____

3. What motivates our market? (Check all that apply then rank in order of importance)

☐ Quality of Design ☐ Quality of Interior Finishes ☐ Streetscape

☐ Master Bedroom ☐ Master Bath ☐ Special Rooms _____

Kitchen: ☐ Brand-name appliances _____

☐ Countertops _____ ☐ Cabinets _____

Location: ☐ Size of lot ☐ Landscaping ☐ Scenic view ☐ Golf Course

Pricing: ☐ Low initial cost ☐ Value (bang for the buck) ☐ Easy financing

4. What is our message? List 10 benefits we provide our buyers, then rank in order of importance.

1. _____
2. _____
3. _____
4. _____
5. _____
6. _____
7. _____
8. _____
9. _____
10. _____

5. How do we reach our market? (check all that apply and describe program. Rate effectiveness of each marketing program.).

☐ Direct Mail (list direct mail materials/frequency of mailing/ size of mailing/ targeted zips

_____

_____

_____

_____

☐ Media Advertising (List publications/frequency size)

_____

_____

_____

_____

☐ On-site Marketing (signage, on-site information center)

_____

☐ Past Customer Referrals (describe program for generating new referrals)

_____

☐ Professional Referrals (list professions (realtors, suppliers, designers, etc. that recommend you))

_____

_____

☐ Parade of Homes/Model Homes (number/cost/effectiveness)

_____

_____

_____

☐ Marketing Materials (List marketing materials and rate effectiveness of each)
    ☐ 4 page positioning brochure (builder story) _____
    ☐ 2 page benefit brochure _____
    ☐ Post-card mailer _____
    ☐ Modular information pieces
       (featuring models and developments) _____
    ☐ Full page ad (8 ½ x 11) _____
    ☐ Half page ad _____

6. Who is our competition? List your top three competitors. What are their major strengths/weaknesses? How can you use those strengths/weaknesses to improve the effectiveness of your own marketing efforts? (See competition shopping form)

1. _____
   _____
   _____
   _____
2. _____
   _____
3. _____
   _____
   _____
   _____

7. Goals. Based on your evaluation of the above, what steps do you need to take to improve your business effectiveness. List ten goals, and then rank in order of importance. Determine implementation dates for each goal.

Goal	Rank	Implementation Date
1. _____	_____	_____
2. _____	_____	_____
3. _____	_____	_____
4. _____	_____	_____
5. _____	_____	_____
6. _____	_____	_____
7. _____	_____	_____
8. _____	_____	_____
9. _____	_____	_____
10. _____	_____	_____

ABC BUILDERS
123 Main Street
Hometown, MD 21234
(301) 231-1234

The purpose of the competitive shopping guide is to find ways to increase your sales and profits. It is used to FIND ADVANTAGES of your product, which you can use to sell, and FIND WEAKNESSES of your product which you can fix or find ways to manage. It will help you be prepared to overcome objections, and to avoid problems in the future. It can also be used to VALIDATE your pricing – too low, too high, or just right.

In filling out this form, try to find a model in your own product line that most closely corresponds to your competitor's model. Then compare the two. Be very hard-nosed and realistic in this comparison. This is NOT to make you feel good about your product, but rather to find SPECIFIC ideas you can use to improve your product.

1. Competitor _____
2. Price Range _____

3. Overall Impressions (rate on a scale of 1-10)
    - A. Product _____
    - B. Presentation _____
    - C. Process _____
    - D. Message/Benefits _____

CATEGORY	THEIR HOUSE	OUR HOUSE	DIFFERENCES Advantage/Disadv.	SIGNIFICANCE of Difference (1-10)
Price				
Location/ neighborhood				
Lot (size, shape, Build ability)				
Exterior Impact				
Landscaping				
Garage (#cars)				
Utilities (water/ Well/septic/gas)				
House size (SF actual)				
Perceived size				

CATEGORY	THEIR HOUSE	OUR HOUSE	DIFFERENCES Advantage/Disadv.	SIGNIFICANCE of Difference (1-10)
Foyer/Entrance				
Volume				
Public Spaces LR/DR/Great Rm				
Study/Library				
# Bedrooms				
MBR size				
MBR bath				
MBR closets				
MBR amenities Fireplace, sitting area, bookcase				
Other bedrooms				
Moldings				
Floorings				
Lighting				
Built-ins/bars				
Natural light/ Windows				
Stairs				
Details				

ABOUT THE AUTHOR

Al Trellis is a custom builder in Columbia, Maryland, and the "Ask Al" columnist in *Builder* magazine. A frequent convention speaker and seminar leader, he is cofounder of Home Builders Network, which provides educational and consulting services to the construction industry.